This innovative book challenges the perceived view (based largely on long observation of artificially fed chimpanzees in Gombe and Mahale National Parks, Tanzania) of the typical social behavior of chimpanzees as aggressive, dominance seeking, and fiercely territorial. In polar opposition, all reports from 'naturalistic' (nonfeeding) field studies are of nonaggressive chimpanzees living peacefully in nonhierarchical groups, on home ranges open to all. These reports have been ignored and downgraded by most of the scientific community.

By utilizing the data from these studies the author is able to construct a model of an egalitarian form of social organization, based on a fluid role relationship of mutual dependence between many charismatic chimpanzees of both sexes and other more dependent members. This highly and necessarily positive mutual dependence system is characteristic of both (undisturbed) chimpanzees and (undisturbed) humans who live by the 'immediate-return' foraging system.

The egalitarians – human and chimpanzee

An anthropological view of social organization

THE EGALITARIANS – HUMAN AND CHIMPANZEE

An anthropological view of social organization

Margaret Power

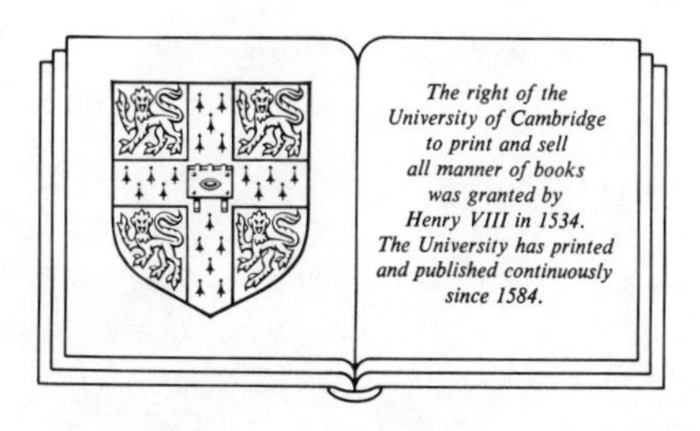

CAMBRIDGE UNIVERSITY PRESS

Cambridge

New York   Port Chester   Melbourne   Sydney

Published by the Press Syndicate of the University of Cambridge
The Pitt Building, Trumpington Street, Cambridge CB2 1RP
40 West 20th Street, New York, NY 10011–4211, USA
10 Stamford Road, Oakleigh, Melbourne 3166, Australia

First published 1991

Printed in Great Britain at the University Press, Cambridge

British Library cataloguing in publication data

Power, Margaret
The egalitarians – human and chimpanzee: an anthropological view
of social organization.
1. Man. Social behaviour II. Chimpanzees. Social behaviour
I. Title
302

Library of Congress cataloguing in publication data

Power, Margaret, 1920–
The egalitarians – human and chimpanzee: an anthropological view
of social organization / by Margaret Power.
p. cm.
Includes index.
ISBN 0 521 40016 3
1. Sociobiology. 2. Human behavior. 3. Primates – Behavior.
4. Social behavior in animals. 5. Social structure. 6. Behavior evolution.
I. Title.
GN365.9.P68 1991
304.5 – dc20 90-48014 CIP

ISBN 0 521 40016 3

Dedicated to the memory of my husband, John Power –
and to mature students everywhere

CONTENTS

Foreword by A. Montagu *page* xiii
Acknowledgments xvi

Part 1: Methods and prefatory explanations **1**
Introduction 1
Attributes shared by humans and chimpanzees 7
Positive and negative behavior and emotive tone 7
The mutual dependence system: a model 8
Methods 11
Preferred use of early studies and methodological tools not currently in favor 14
Direct and indirect competition 15
Charisma and dependence 16
The functional approach 17
Use of the term foragers 17
The (sometimes) harmless people, and the (normally) friendly chimpanzee 18
The naturalistic studies: practitioners and methods 21
Feeding methods: the Gombe Research Center 28
Feeding methods: the Mahale Research Center 30

Part 2: The human foragers **37**
General anthropological understandings 37
The immediate-return foraging system 39
Territoriality 43
Social kinship 45
Leadership 46
The foragers' sanctions 48

Part 3: The changing social order **51**
Kortlandt's hypothesis 51

Mobile and sedentary groups 55
The social climate of wild groups 57
The territoriality of wild chimpanzees 60
The social climate of the provisioned groups 67
The recent territoriality of the Gombe and Mahale apes 70
The rise of despots 74
Sexual relations among wild chimpanzees 77
Sexual relations in the Gombe group, post-1965 79
A 'gentle crucible': the infant experience in wild chimpanzee groups 84
The Gombe infant experience 88
Infanticide and cannibalism at Gombe and Mahale 95
Persecution of the vulnerable 103
Prolonged dependency, delayed social maturation 105
Migration among wild chimpanzees 114
Apprentice into adolescent 117
Solidifying alliances 122

Part 4: The behavior of wild and provisioned groups: a theoretical analysis **125**
Foraging theory: when, and when not, to defend territory 125
Costs and benefits of recent behaviors of the Gombe and Mahale apes 127
The foraging strategy of wild chimpanzees 133
Frustration theory 136
Barker *et al.*'s experiment 137
Behavior without a goal 139
Competition theory 143

Part 5: The mutual dependence system **151**
The nature of authority 152
The parental roles 155
Variability of temperament 161
The nature of true leadership 167
Nishida's and de Waal's insights 168
Traits of charismatic leaders 170
The nature of the follower role 171
The nature of deference 172
The charismatic–dependent role relationship 174
A generalization of the parent–child relationship 174
Noninterference mutualism 178
The means of social control 179
The independent conformists 179
The use of sanctions by chimpanzees 181
The positive sanctions; reward and goal 186
The structure of attention 188

Part 6: The egalitarian chimpanzees **195**
In search of the female 196
Achievement of adult rank 203
Acceptance through grooming: an adult male–adolescent device 209
Acceptance through aunting: a female–female way 214
Acceptance through copulation: an opposite-sex way 218
The social function of the carnival reunions 224
The function of the mobile and sedentary groups 232
The role of charismatic elders 234

Part 7: Probabilities, possibilities and half-heard whispers **239**
The adaptive value of negative behavior: a possibility 239
The chimpanzee hierarchy 244
Further suggestions and comments 248
The Rousseauean foragers: human and chimpanzee 248
A mammalian system? 253
Focus for the future: pro-social behavior 253

Notes 255
References 266
Index 281

FOREWORD

Every so often there appears a work of synthesis which is so original and creative that it clarifies and systematizes a whole field of observation or knowledge that had hitherto been adventitious and chaotic. Such were the works of Rudolf Virchow, whose *Cellular-Pathologie* (1858) not only founded the science of pathology, but constituted the fundamental contribution to the rise of modern medicine, as was also Claude Bernard's *Introduction à l'étude de la Médecine Experimentale* (1865), which put physiology on a firm scientific basis as well as the founding of regulatory biology and its relation to human health. What these books, each in its own way, did for the growth and development of medicine and human health, I am convinced that Mrs Power's admirable book will do for the growth and development of our better understanding of the dynamics of social life not only among chimpanzees, but for all primates, including humans. For what the author has done is to make a microscopic examination of the fieldwork of numerous independent investigators who have studied chimpanzees under natural or artificial conditions. With great acumen she has seen that even under natural conditions, the conditions may not be as 'natural' as most investigators and their interpreters and readers may have thought, and that various factors under such conditions have led to a variety of inferences which have in fact been erroneous. Mrs Power's analysis of these factors for the first time really makes it possible to train future students of animal behavior, and more particularly of primate behavior, including humans, in the delicate methodology of fieldwork. At the same time it makes it possible to understand, and to avoid, the erroneous conclusions that have been drawn from the studies of earlier investigators, one of which especially exercised an enormous and obfuscating influence upon almost three generations of social scientists, anthropologists, explicators of human behavior of various sorts, and others – all based on observations made on a colony of baboons at the London zoo. The book which reported these observations was Solly Zuckerman's *The Social Life of Monkeys and Apes*, published in 1932, and republished in 1981, with a postscript in which the author defends his earlier work. It is a spirited riposte and is well worth reading, but, while recognizing the nature of the criticism to which his conclusions have been subjected Zuckerman fails to understand that, however valid they may have been for animals living under the crowded conditions of the zoo colony, they are wholly inapplicable to the behavior of animals living under natural conditions.

Even under natural conditions, when environmental conditions are varied such changes will affect the behavior of the animals studied, a fact which is especially true of primates. This is a point which Mrs Power abundantly and convincingly shows has often had a major effect upon the structure and functions of a whole chimpanzee society. She shows how, by the introduction of what may appear to us as a simple change in a chimpanzee society, it may change from a peaceful highly cooperative community into a highly aggressive and virulent one. When such destructive behavior by chimpanzees was first reported it was immediately interpreted in the press to mean that our 'bad habits' are derived from 'our animal ancestry' (Editorial, *The New York Times*, 2 May 1978, p. 34). Such a conclusion would have pleased the 'innate aggressionists' such as Konrad Lorenz, Robert Ardrey, and many other believers in 'innate depravity', but the facts are quite otherwise than those reported in the press. Mrs Power's discussion of the causes of the sudden development of highly aggressive behavior in Jane Goodall's chimpanzees, is brilliantly illuminating, for not only does it correct endemic errors by setting forth the facts, but also points to certain principles which all who investigate, write about, or are otherwise interested in animal and human behavior would do well to make part of their intellectual equipment.

Altogether Margaret Power has produced an account of the social life of humankind's closest relative, the chimpanzee living under natural conditions. She has illuminatingly shown how, under varying conditions, and especially the varying conditions which humans have introduced into their lives, chimpanzees will undergo profound changes in their social lives, very much resembling the kinds of change which humans have undergone in the course of social evolution. The book will, therefore, be indispensable to all students of human society as well as the behavior of humans under different conditions. As such, Mrs Power's book constitutes a gold mine of new ways of looking at old problems which will be of the most exciting and heuristic value to all who may be interested in the relation between the nature of the structure, functions, and environmental changes which condition societies such as those of chimpanzees and human beings.

Mrs Power writes sensitively and as a scientist intent on teasing out the factors which affect the individual behavior and societies of chimpanzees, to correct many endemic errors concerning such societies, and to set the record straight concerning the lives of chimpanzees under changes in the conditions of life which may result in extreme changes in behavior as well as in physiologically striking effects which may in turn produce marked adaptive changes in the structure and functions of the societies in which they occur. In so doing, Mrs Power has succeeded in overturning many entrenched beliefs concerning our close relative the chimpanzee, an accomplishment which is bound to have a most constructive and salubrious influence in bringing about a reordering of scientific as well as popular thinking concerning not only the chimpanzee but also ourselves and our remote ancestors.

All students of human and animal nature owe a great debt to Mrs Power for the brilliant light she has thrown upon so many complex problems which over the course of the years so many hundreds of workers, thinkers, and specialists have contributed to that body of knowledge known as primatology. Our author has for the first time put together the relevant facts so that one can now really see the wood for the trees, and make sense out of what has up to now prevented us from clearly seeing how it is that each tree is affected by innumerable different influences, which together constitute the wood and the varying changes it exhibits under varying conditions.

It is a book that is bound to become a classic, one that will be read with interest and profit for many years.

Ashley Montagu

Princeton, New Jersey

ACKNOWLEDGMENTS

Throughout the 12 years that I have spent developing this work, many people have sustained me through their interest, enthusiasm, comments and criticism.

I am particularly grateful to Vernon Reynolds and Michael Chance, who were the first to see in a very undeveloped idea and an exceedingly rough early draft something worthy of further exploration. Their interest resulted in a working visit to the Department of Biological Anthropology at Oxford University in 1981, and thanks are due to the British Council for a grant which made that visit possible. Michael Chance not only has kept up a voluminous correspondence pertaining to my developing mutual dependence concept but also made a trip to Canada in order to discuss at length with me my ongoing work. Discussions with both researchers, particularly in the early years, helped me to clarify my ideas and find my direction.

Geza Teleki has been more than generous, giving of his time and knowledge by reading early drafts and offering much appreciated advice and constructive criticism, as also have been Thelma Rowell, Claire and William Russell and Pamela Asquith. Robert Hinde offered encouragement and much needed criticism in the early years, both of which were equally helpful.

I owe special thanks to Ashley Montagu, whose help over the past few years has meant so much. Not only has his work been a major influence on my thinking for several decades, but he continues to give generously of time gleaned from his many academic and public commitments to assist a previously unknown disciple.

Ian Whitaker, my MA supervisor, has been an unfailing source of support and encouragement since my undergraduate days. His belief in my abilities, and readiness to accept my independent, usually unorthodox, approach to anthropological questions, gave me belief in my abilities and freed me to follow my own path toward understanding the primate, both human and nonhuman.

This book grew out of a Woman's Studies course taken after I had graduated. I know that had I not taken that 'eye-opening' course this book would not have been written.

I thank the staff of the University of British Columbia library, especially Ann Yandle, Head of the Special Collections department, who always found space for me to work, and senior supervisors Erik De Bruijn, Tom Shorthouse and Alan Soroka, who gave me time to write. Robert (Bert) Forrest deserves special thanks for his unfailing patience in retyping the study 'just one more time,' adding 'just one more insertion,' and for his enthusiastic comments and suggestions, all of which I appreciate. Thanks also to Alan Johnstone, proof-reader extraordinaire.

I also thank my children, Tony, Laura and Rosemary, for their warm pride in (and occasionally patient resignation with) an absentminded and often absent mother. Their supportive attitude smoothed my path. The contribution of my late husband, John Power, cannot be measured. It was limitless.

I recognize the great debt that I owe to the researchers whose field studies and conclusions are the basis for this book. Although I disagree with many of their formulations, I thank them for the meticulous honesty of their reporting, which made this study possible.

Dr Jane Goodall has been involved in field studies of free-living chimpanzees in Gombe National Park, Tanzania, for more than three decades. I have greatly admired Dr Goodall for her courage, dedication, tenacity, integrity, openness and meticulous attention to detail for approximately the same number of years.

A main focus of this book is to demonstrate, through reference to the publications, that the quality and quantity of social change experienced by the most-studied Gombe group, the Kasakela community, over these 30 years, is greatly underestimated. Indeed, the quite usual, widespread acceptance of the behavior of this group as being the normal behavior of undisturbed, wild champanzees, has dangerously skewed our understandings.

Dr Goodall is by far the most frequently published and primary source of data on the social behavior of the Gombe chimpanzees.

Accordingly, I cite her data and interpretations far more than I do those of any other Gombe researcher. This may give an impression that I am especially critical of Dr Goodall's work. This is not the case. Indeed, I am pleased to find that I retain my deep and sincere admiration for Dr Goodall, intact and unweakened, despite my different perspective.

Margaret Power
2–1786 Esguimalt Avenue
West Vancouver
British Columbia
Canada V7V 1R8

'Whatever happens to the beasts,
soon happens to man.
All things are connected.'

Chief Seathl, 1854.
Indian Orator,
for whom Seattle, Washington,
is named.
Taken down and translated
by a young Seattle pioneer,
Dr Smith.

PART

1

Methods and prefatory explanations

Introduction

Despite more than 30 years of study of free-living chimpanzees in their African habitat, there is no firm agreement as to the social organization of this species. Both 'naturalistic' (unobtrusive) and 'provisioning' (artificially feeding) methods of field study have been used, with sharply differing results. Independent field researchers, using various naturalistic methods report – quite separately but with striking unanimity – peaceful, open groups of nonaggressive chimpanzees without signs of any dominance hierarchy, enforced territoriality or single leaders.

At the long-established, permanent centers for the study of free chimpanzees in Gombe and Mahale National Parks, Tanzania, artificial feeding methods have been used. Researchers at both centers – also quite separately – report many strongly similar aspects of behavior and social organization of the provisioned apes that are opposite to reports from naturalistic studies. The artificially fed Gombe and Mahale apes are extremely aggressive, dominance seeking, directly competitive and fiercely territorial.

Habituating the apes to the presence of humans through provisioning facilitates excellent, lengthy observation opportunities, an intimate knowledge of the chimpanzees as interacting individuals, and a large pool of data. Naturalistic studies do not have these advantages. Consequently, not only are our current understandings of chimpanzee social behavior and organization based very largely on the Gombe and Mahale studies but also there is a tendency to assume that the noncontinuous naturalistic studies yield few or no data on the social behavior and organization of chimpanzees. Washburn (1980:258) expresses the view of many when he asserts that 'the beginning of reliable studies' on the natural behavior of chimpanzees may be marked by 'Jane Goodall's (*1968*) long-continued investigations' (my emphasis). The sharply different 'naturalistic' reports are assumed to stem 'from the conditions of the early field work and the human desires (for peace and harmony) of the 1960s.' Washburn also dismisses Goodall's first 8 years of field study, pointing out that the results of post-1968 Gombe studies are 'facts of an entirely different order' from earlier perceptions of 'the friendly chimpanzee.' I differ in that I find all of Dr Goodall's publications highly valuable because her pre- and post-1968 studies are, as Washburn suggests, facts of very different (social) orders. In this study, I rely heavily on both her naturalistic (pre-1968 (or, as I shall argue, pre-1965)) studies and her post-1965 provisioning studies.

The current consensus based on post-1965 Gombe (and post-1968 Mahale) studies, is that chimpanzee society normally and generally is organized around a core group of closely related males who aggressively restrict the access of others to their territory, its resources and the breeding females. The complete acceptance of aggression as a normal part of chimpanzee social life is testified to in Goodall's (1986*b*) carefully documented, important monograph, *The Chimpanzees of Gombe*. Goodall (1986*b*:3, 55) explains that, as much as is possible, her book is based on data collected 'since 1975,' because the data from earlier years presents a 'very different picture of the Gombe chimpanzees' as being 'far more peaceable than humans.' These early data give a wrong impression, Goodall believes, in that 'aggression is part of the complex network of social relationships within the chimpanzee community and, along with the other patterns of agonistic and friendly behavior, it plays its role in structuring chimpanzee society' (Goodall 1986*b*:353). She suggests that 'it is the interplay between those two opposing forces, aggressive hostility and punishment on the one side and close and enduring friendly bonds on the other, which has led to the unique social organization that we label a community' (Goodall 1986*b*:356). A main focus of my study is to show that the juxtaposition of hostile and friendly relations as 'equally powerful forces' (Goodall 1986*b*:356) now so evident among the long studied Gombe chimpanzees is not characteristic of the organization of chimpanzee groups generally, but a distortion caused by abnormal (human-imposed) stress.

The Mahale and Gombe observers recognize that the years of artificial feeding have caused some changes in the behavior and social organization of the fed groups; but the consensus of scholarly opinion seems to be that any changes resulting from provisioning are minor, and quantitative not qualitative. It is assumed that the behavior of the fed apes is somewhat more aggressive than would be typical in a natural feeding situation; and that, while provisioning does induce some frustration because all apes are not able to obtain some of the small ration of bait foods, the recent research results accurately reflect the intrinsic qualities and behavior of chimpanzees in general. However, Maier's (1961) hypothesis is that frustration causes a distinct behavioral change in the condition of an organism; that the normal, constructive, problem-solving (positive) processes are replaced by a different set of nonconstructive or destructive (negative) behavior mechanisms.

For these reasons, one goal of this book is to show, through the use of published evidence from both naturalistic and provisioning field studies, that (quite without such intent) the artificial feeding used at both the Gombe and Mahale Research Centers deeply frustrated the chimpanzees, which precipitated extensive, qualitative change in their behavior and organization. Hence, those using naturalistic and provisioning methods *are* viewing a different kind of evidence. If we wish to understand the full social potential of chimpanzees, *neither set of evidence can be ignored*. Moreover, if early Gombe facts are not the same as recent Gombe facts, it would not do to assume that the same theoretical model will equally well explain both sets of evidence; or if a trusted, long relied on paradigm fails to explain both, then on that basis, one or the other set of evidence is poor, weak or erroneous. It may be that different structural models are required, for what are widely differing sets of evidence. These are the presumptions on which this study proceeded.

Quite unexpectedly, a third theme interjected itself, early in the course of my research. On the basis of studies of six small, still existing, gathering–hunting (foraging) peoples, Dr James Woodburn (1982) has produced an anthropological model of a very simple, highly egalitarian foraging system that he refers to as the 'immediate-return' system. Very suddenly, I realized that despite their not being human, undisturbed (wild) chimpanzee groups meet all of Woodburn's criteria for being foraging societies living by the immediate-return system. This concurrence suggested that the model of *mutual dependence* organization, which I was in the process of developing through a restudy of the publications from naturalistic field studies of chimpanzees, might also be used to further our understanding of the fundamental principles underlying the social organization of these human foragers; and anthropological understandings regarding these human societies might further our understanding of chimpanzee social organization. What is the most simple full form of human organization, Woodburn's immediate-return foraging system, is more broadly a primate model.

Hinde (1986:413) warns that drawing direct parallels in behavior between human and nonhuman primate species is dangerous, because doing so 'could produce a very different perspective.' It does do so, but there is much direct evidence from the naturalistic field studies to support the new view. At the same time, I wish to make it abundantly clear that it is not suggested that their sharing the same socioecological adaptation to the problems of a foraging way of life implies any blurring of the two species, *Homo sapiens* and *Pan troglodytes*. *Neither is a semi-human species*. All that I suggest is that these two species of primates arrived at the same organizational solution to similar socioecological problems.

Quite correctly, Reynolds (1970) warns that we must be very careful in extending human-based concepts to a nonhuman species, because seemingly similar behavior patterns can have very different functions in the lives of the two species. They do not necessarily imply identical underlying causes or motivations. However, I write of a fully developed *system* of social organization which both chimpanzee and human foragers follow, an adapted way of life in which positive social relations form the social structure. While we cannot know if the motivation is the same, patterns that are part of this form of social organization should function similarly in the societies of both species. The nature of this type of social organization itself generates the necessity of certain behaviors, constraints and principles in any group so organized. Many aspects of social behavior, interactions, roles and relationships have become essential functional and structural parts of the social organization. Accordingly, the fundamental social behavior of species living by the 'immediate-return' foraging system will be the same, be the actors human or nonhuman. (The immediate-return system is explained in Part 2).

Hinde and Stevenson-Hinde (1976) suggest judicious use of many fundamental, human-based concepts and principles to do with relationships, from many disciplines, may be enlightening in understanding nonhuman primate behavior and organization, and also that some animal studies may provide a testing ground for concepts useful for study of the more complex human case. Following this suggestion, I draw on a number of human-based theories regarding affiliation, competition, leadership, dependence and so on, and find them illuminating toward understanding chimpanzee behavior and relationships. In developing the argument in this volume, the general understanding that humans and chimpanzees share the same basic emotive spectrum, used in the same way, is highly important. Certain human-based psychological theories are found to be useful tools.

This eclectic work offers a new theoretical approach to old problems of interdependence, attraction to the group, leader and follower status and roles, and so on. The intent is to make clear a new perspective to a wide variety of readers. It would be unwise to assume that the reader, encountering a new argument that is opposite to established and popular current views, would be willing to leap from one peak idea to the next. This is unfamiliar territory, and, to know it, we must explore it thoroughly.

Because of the new perspective proposed in this study, some explanations thought necessary to make clear my methods are offered in Part 1. Part 2 enlarges on Woodburn's concept of an immediate-return system, and outlines anthropological understandings regarding human foraging (gathering–hunting) societies who live by the immediate-return system.

In Part 3, chimpanzee behavior as reported by naturalistic and provisioning researchers is compared, and an attempt is made to establish a chronological history of the social change that has taken place among the Gombe and Mahale apes since artificial feeding was begun. By considering the Gombe and Mahale findings and the reports from the naturalistic studies in terms of frustration-aggression, competition and foraging theories, it is argued in Part 4 that the artificial feeding is a catalyst for a high degree of qualitative, negative social change among the provisioned groups, which spread in ripple fashion to distort all aspects of the adapted social order.

The form and structure chimpanzee society takes under normal conditions is the subject of Parts 5 and 6. A close reading of the publications from naturalistic studies reveals that there is sufficient evidence from which to construct a preliminary theoretical model of an egalitarian form of social organization based on *mutual dependence,* which is opposite in character (nature, or quality) to current understanding. In the final section of this volume, Part 7, it is tentatively suggested that the adapted form of social organization of chimpanzees is a dominance hierarchy, but that according to ecological circumstances (ideal or crisis) the hierarchy may be, in form, an almost invisible, seeming unstructured 'correlational' type or a more visible, authoritarian, structured rank order.

This is a library study, a synthesis, based on many others' years of difficult field work. The contribution of the field researchers is enormous, and my debt to them all is equally immense. While in this volume the post-1965 Gombe and post-1968 Mahale data are rejected as being representative of the customary behavior and organization of wild chimpanzees, this specific rejection does not imply discard of these important data. To the contrary, these data are of the *utmost* importance. They are detailed, long-term, utterly invaluable records of the spread of social change in two wild groups, unintentionally set in motion by restrictive provisioning methods. As a perceptive scientist suggested to Kuhn (1970:85), introducing a new paradigm does not necessarily involve discard of the established model; sometimes it is a matter of 'picking up the other end of the stick, ... of handling the same bundle of data as before, but placing them in a new system of relations with one another by giving them a different framework.' This is very much the case, in two senses, in my suggestion of this new view of chimpanzee social organization. The first, as suggested, is a reinterpretation of published analyses of Gombe and Mahale data. But another way of picking up the other end of the stick is by using the neglected, very different reports from naturalistic field studies of wild (not artificially fed) chimpanzees. Only together do the two very different sets of evidence reveal the possible social polarity of chimpanzee behavior and organization. Both offer understandings of enormous value to our own, increasingly stressed, human species.

I hope that the proffered model of a system of *mutual dependence* will be critically appraised firstly on the basis of the coherence and logical consistency of the argument, and only secondarily on its data base. The *mutual dependence* model is offered as a *preliminary* model of one previously unrecognized form of primate organization, a starting place for further studies and development. Clearly, a great deal more work must be done before we can reach any firm conclusions. At the same time, our need to understand the adapted social organization of our nearest nonhuman relation, the chimpanzee, is urgent. We must therefore work with what we have and then, through further research, verify – or disprove – the resulting hypotheses.

Attributes shared by humans and chimpanzees

Chimpanzee social behavior is the most plastic and humanlike among that of existing nonhuman primates. Recent research makes it increasingly evident that the chimpanzees have close genetic, morphological, physiological and behavioral affinities to humans.

The fundamental emotions are an innate aspect of human nature which is assumed to be not only transcultural, but also trans-specific (Izard 1972). Hebb (1946) suggests that not only may the same processes be used to recognize emotions in chimpanzees and humans, but that our recognition may be more accurate in viewing the apes. Their expression of emotion is usually more direct and uninhibited. We humans more often wear a mask.

In 1972 and 1973, Gallup carried out experiments with chimpanzees' responses to their mirror images which demonstrated that chimpanzees share with humans self-awareness, a perception of self. To be self-aware is to share a fundamental psychology with humans, Shafton (1976) suggests, not the same mental powers, but the same basic emotive nature and mental processes. Cognitively, of course, humans and apes are worlds apart. At the same time, as a result of his close study and intimate knowledge of the social behavior of the chimpanzees in The Burger's Zoo, Arnhem, The Netherlands, Dr Frans de Waal (1982:42) suggests that we tend to underestimate the sophistication (in terms of lack of lower animal instinctiveness) of the emotion-based responses of chimpanzees. De Waal (1982:51) is convinced that the ability of these animals to use reason and rational thought (defined as being terms used to describe 'the ability to make new combinations of past experiences in order to achieve a goal') is socially almost equal to that of humans, though technically greatly inferior.

Positive and negative behavior and emotive tone

The bipolar nature of emotions is widely recognized. Scientists find it useful to classify emotions as either positive (pleasurable), or negative (unpleasurable), on the basis of their sensory/experiential characteristics. The emotions, or subjective feeling-states known by individuals in their daily lives, may be thought of in terms of the emotive 'tone' subjectively experienced. The *emotional tone* of a society is usually defined as being the product of the separate behavioral responses of large numbers of the population in coping with their social environment.

Izard and Tomkins (1966) suggest that there is a strong connection between affect or emotion, and behavior. 'Phenomenologically positive affect has inherent characteristics that tend to enhance one's sense of well-being and to instigate and sustain approach toward and constructive relations with the object' (Izard and Tomkins 1966:87).

Positive emotive feeling, according to Selye (1974:75), 'can best be described as "love" in its broadest sense.' It includes such qualities as gratitude, respect, trust, admiration, goodwill, friendship and so on – to use Selye's examples. Negative feelings (in the broadest sense, hatred) include anger, distrust, disdain, hostility, jealousy, and the urge for revenge – in short, every quality 'likely to endanger your security by inciting aggressiveness in others who are afraid you might harm them,' Selye adds. Negative feelings disturb and distress, and involve retreat from, or destructive relations with, the involved object.

Scientists do not usually attempt to understand primate societies in terms of the emotive tone of social interactions, behaviors and relationships. However, the terms positive and negative are used throughout this work, to classify and point up the differences in the behavior and emotive tone of the artificially fed Gombe (and Mahale), and the (relatively) undisturbed wild chimpanzee groups. As used in this book, *positive* behaviors, interactions and relationships are those based on positive affect which reinforce (are in accord with) the statuses and roles that are the organizing phenomena of the normal social order. *Negative* behaviors are those which are disruptive or disorganizing, in opposition to the smooth functioning of the adapted social order. In other words, negative behaviors, interactions and relationships act to change a social structure. Positive behaviors act to maintain it. Sharing the same spectrum of emotions and range of emotional expression as humans, chimpanzees can be expected to be peaceful (a positive sensory/experiential characteristic) in some conditions, and aggressive (a negative sensory/experiential characteristic) in others.

The mutual dependence system: a model

Because the concept of a social system of mutual dependence is new, it might be helpful to outline it at this point. (It is developed as fully as possible from the evidence available, as Part 5). The main principles which compose the structure of a mutual dependence order are as follows:

1 The pattern of fission and fusion, operating at subgroup, local group and larger society levels.
2 Open groups ranging familiar, undefended, typically overlapping home ranges, which are local units of a larger interacting, interbreeding population.
3 Indirect competition.
4 A fluid and typically interchanging charismatic leader–dependent follower role relationship, which is one of noninterference mutualism (NIM), i.e. beneficial to the participants without cost to others (Wrangham 1982).
5 Mobile and sedentary forms of subgroups.
6 Individual autonomy or self-direction as a property of the social system.
7 A social system which is extraordinarily egalitarian.

The principle of charismatic leader–dependent follower status/roles permeates all levels of society. This very fluid relationship operates between individuals, charismatic and dependent, mobile and sedentary subgroups, groups and local groups. Group composition changes constantly. Particularly among chimpanzees it would be most unusual and uncharacteristic for exactly the same subgroup to reform.

Because the pattern is for the subgroups to change personnel very frequently and because there are no permanent leaders, foraging and traveling subgroups appear to form randomly, on the basis of mutual congeniality. Although exceedingly fluid in composition, the subgroups are not without structure. Each small, very temporary subgroup is composed of at least one, or perhaps several confident, charismatic individuals – of either sex – and one or usually several, more nervous or less assured, hence more dependent, individuals, also of either sex, who are attracted by the charisma of some confident member of the group whom they choose to follow, for a time. Thus, *the composition of the subgroups constantly changes, but the structural form remains the same* (see Figures 1 and 2).

One important result of this egalitarian mutual dependence system is a high level of peaceful sociality within and between groups. It is argued in Part 4 that the above form of organization is optimal in terms of foraging strategy, in the type of natural habitat to which the common chimpanzee is adapted.

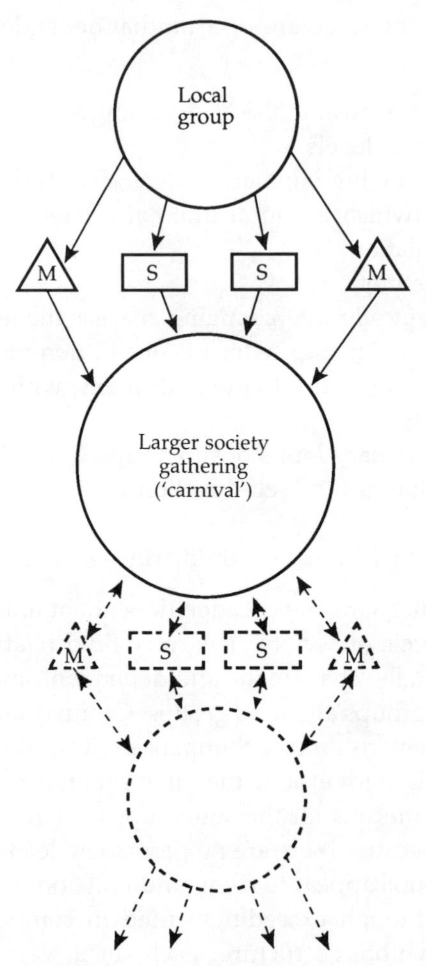

Mobile charismatic-dependent subgroups: numerically, male. *Structurally, the childless, active adults and near-adults of both sexes.*

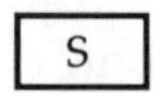

Sedentary charismatic-dependent subgroups: numerically, female. *Structurally, the less active by role or inclination*. Childrearers (mothers), dependent young, the elderly of both sexes.

Figure 1. Structures movement and grouping pattern: the mutual dependence system. Movement in any type of group is two-directional.

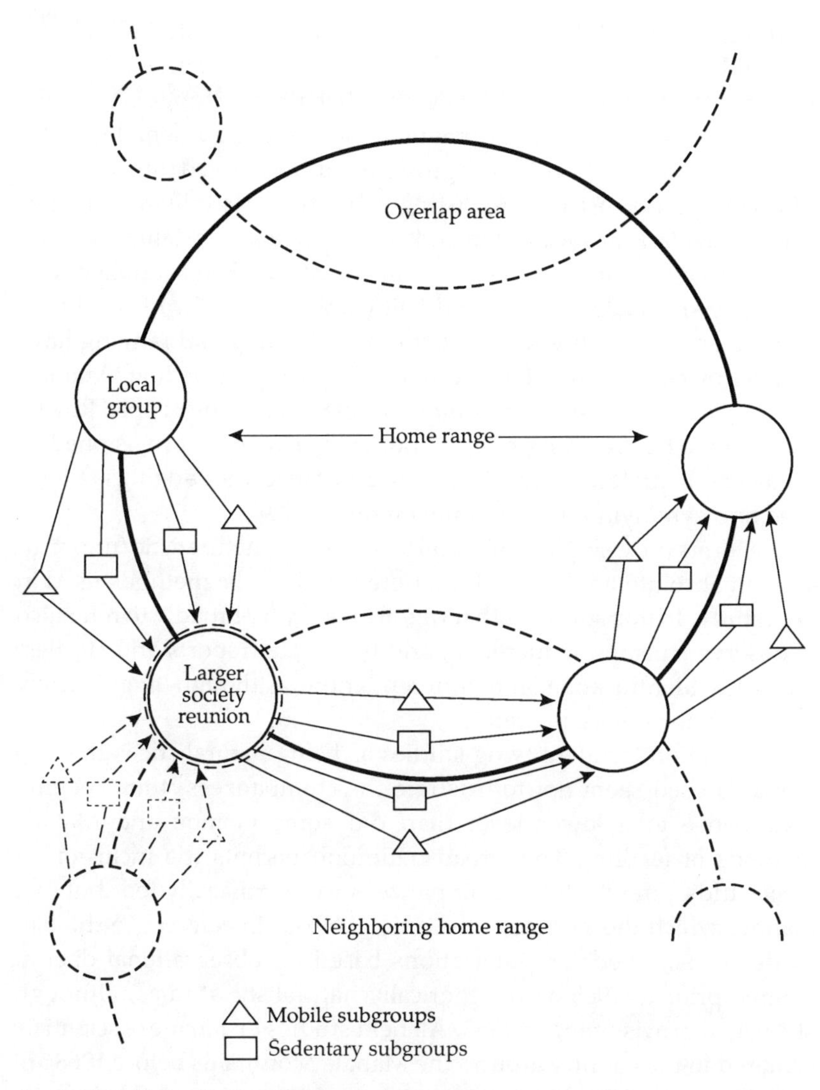

Figure 2. Pattern of movement of the group about its home range. There is some exchange of members when two local groups meet in carnival reunion, usually in the overlap area.

Methods

An explanation of some of the innovative methods used is necessary at this point. In order to develop the argument, I make a number of arbitrary binary distinctions between: (1) *wild* and *provisioned* chimpanzees; (2) *naturalistic* and *provisioning* methods and studies; and (3) *early* and *recent* Gombe and Mahale studies and data. These categories and the terms positive and negative behavior distinguish crucial variables on which the development of the whole argument depends.

The two categories 'wild' and 'provisioned' are derived from Rowell's (1972) recognition of a need to distinguish between free-living primates that are regularly artificially fed and wild primates. She defines as *wild*, undisturbed primates and also animals which have been supplied with only very small amounts of food bait (Rowell 1972:67). I take this to indicate animals whose natural patterns of feeding and ranging have not, in Rowell's judgment, been changed by artificial feeding. Animals which 'are not confined but fed almost all of their food by people' Rowell considers to be 'free-ranging', but not wild. The term 'provisioned' is used here, in preference to free-ranging, because free-ranging is a term used as a synonym for wild in the literature.

The terms provisioning (artificial feeding) and naturalistic (unobtrusive, an absolute minimum of interference) describe methods used in field study. In this study, both terms are used very broadly to refer also to observers using the methods, and their data, reports and studies, for quick identification and differentiation. Although linguistically improper, the device is useful.

My criterion for identifying studies as being naturalistic is that the methods used seem not to frustrate the chimpanzees, thus keeping disturbance to a lower level than did some Gombe and Mahale methods of feeding. This broad definition permits the inclusion of some studies in which the chimpanzees were artificially fed, but in a manner which did not block the animals from free access to the bait foods. Thus, Goodall's publications based on observational data at Gombe prior to 1965 are categorically naturalistic studies, although she began provisioning in 1963. All field studies by Japanese scientists centered in the country around the Mahale Mountains before 1968 are naturalistic studies, although they too sometimes provided bait food.

Those chimpanzee groups inhabiting Gombe National Park which have been artificially fed since 1963 are categorized as being *wild* chimpanzees studied by *naturalistic* methods *up to 1965*, when a restrictive human-controlled feeding method was introduced. Studies of these animals done prior to that date are designated *early* studies. *After 1965*, these same apes are defined as being *provisioned* chimpanzees, and the studies are designated *recent* or *provisioning* studies.

Similarly, the *early* studies of the two habituated groups (M and K) in Mahale National Park are of *wild* apes studied by *naturalistic* methods *to 1968*, when, there too, a new feeding method was introduced that entailed human control over access of the chimpanzees to present but withheld bait food. *Recent* Mahale studies are those carried out *after 1968*: after that date the Mahale apes too must be designated *provisioned* chimpanzees. All other field studies referred to in this volume are categorically naturalistic studies.

Almost all of the many studies carried out at both permanent centers over the past two decades have been of two small habituated groups (of five or six local wild groups living in each area) which regularly seek the bait foods (Goodall *et al.* 1979; Nishida 1979). In fact, by 1971 at Gombe and 1975 at Mahale, one of the two studied groups had disappeared. They are thought – at Gombe, known – to have been exterminated by the other habituated group. Thus, all major studies carried out in both areas since those dates have been on *one* provisioned group. We must reconsider the wisdom of placing quite so much weight on what is an extremely small, albeit long-studied, sample, as we have done, to this date.

When the *Gombe* or the *Mahale* chimpanzees or groups are mentioned, it is these artificially fed groups that are indicated. It is an important point, which will be expanded later, that no researcher other than Goodall and her assistants studied the Gombe apes until 1967. Therefore, literally *all* of the Gombe data and studies, with the exception of Goodall's early 5 years (1960–5) of naturalistic observation (and the early Mahale studies carried out before 1968), are *recent* information and studies, based on observation of frustrated, provisioned chimpanzees.

De Waal's studies are based on his and his colleagues many years of close observation of the social behavior of captive chimpanzees in The Burger's Zoo, Arnhem (de Waal and van Roosmalen 1979; de Waal and Hoekstra 1980; de Waal 1982). Although these apes are kept in superior zoo conditions, there are many unnatural stresses in even these fine zoo conditions.[1] Despite this, there is high value in these studies, in that the zoo situation permits more intensive, close, day by day study of the captive animals than is possible in the wild. As Goodall (1986*b*:583) points out, observations of the Arnhem group have contributed meaningfully to our understanding of the upper limits of (chimpanzee) social intelligence. Studies of these captive apes are not used herein to argue what wild chimpanzees do, but they are useful to illustrate what chimpanzees can or may do.

Because of a general overreliance on the Gombe and Mahale studies, and on the paradigm of an aggression-based dominance rank order with the result of fixation of much scientific attention on the behavior of male chimpanzees, I am forced to rely more heavily than is ideal on the shorter, less continuous, hence more impressionistic naturalistic field studies for evidence from which to build a model of the adapted social organization of chimpanzees. (The unobtrusive, 'naturalistic' field observers also attempted to explain the highly positive interactions and behaviors that they observed in terms of the trusted dominance model. My solution to this particular problem is acceptance of their data, but rejection of their analyses, which necessitated a reinterpretation of their understandings).

Preferred use of early studies and methodological tools not currently in favor

Also found useful is a preferential use of data from earlier, rather than more recent, studies of the immediate-return gathering–hunting (foraging band) peoples. My reason is that extensive disruption of the traditional way of life has been the common experience of foragers everywhere, as they came into contact with other peoples with more structured forms of government.

For example, among the most remote foragers remaining are the !Kung San (people) of the Kalahari Desert. It was 1960 before direct European colonial governmental administration of the !Kung was introduced (Lee 1979). Since then, the pace of change has accelerated, and by the 1970s the !Kung were being swept away from the free nomadic foraging life. These once independent hunter–gatherers were pushed into becoming farmers and herders, migrant laborers, soldiers in the South African Army, and displaced dwellers on reservations.

At a 1965 symposium on 'Man the Hunter,' which brought together then-current research on the gathering and hunting peoples, some participants advocated disregarding early reports of nonaggressive foragers in favor of more recent reports of aggression and competition among disturbed groups. Some – in striking parallel to the view of many observers of the recent behaviors of the interfered-with Gombe and Mahale apes – assumed that the difference between recent and early reports is the result of longer, more systematic methods of research. Little weight was given to the fact that recent reports of aggressive patterns of war, competition, territoriality and such, among groups still designated as foragers, are of people who have been forced to take up new ways of subsistence, living only marginally in their ancient foraging way.

Levi-Strauss (1968:352) warned the scientists at the 1965 symposium against dismissing the earlier work of their predecessors. He pointed out that 'we cannot be sure we are actually observing the same kind of evidence.' (It has already been suggested that neither are the Gombe and Mahale researchers observing the same kind or set of evidence, as are the naturalistic observers of wild chimpanzees).

Hence, when what the reader might consider as outmoded findings regarding the behavior of human foragers and of chimpanzees are cited in this volume, *it is because the recent understandings are based largely on the studies of provisioned chimpanzees and human foraging groups under the pressure of forced, rapid social change*. The scant, hard-won data from the naturalistic field studies are all we have of the behavior of free chimpanzees following the normal way of life to which they are adapted. Similarly, data from early field studies of the few immediate-return foraging groups are all we have of what may have been the fundamental system of human organization.

Paradigms, theories, hypotheses, concepts and definitions are important tools of science. Constructs currently subscribed to are those that are found to be the most useful toward understanding the subject or phenomenon under study, that offer solutions to the scientific practitioners. In our efforts to understand the adapted social organization of chimpanzees, the tools thought to be appropriate are often those found to be useful in explaining the recent behavior of the Gombe and Mahale apes.

These are not always the best tools for understanding wild chimpanzees generally. Accordingly, sometimes concepts not currently in favor are deliberately chosen, because they are found to be more useful tools for analysis of the data from the naturalistic field studies and development of the mutual dependence model.

Direct and indirect competition

Competition is defined as being simply 'the simultaneous seeking of an essential resource of the environment that is in limited supply' (Mayr 1970:415). This definition does not necessarily imply direct (or interference) competition in which two or more individuals compete against each other for the same resource object or objects. A society fissioning to forage for the same kinds of food, in the same general area, is an example of *indirect* competition.

The distinction between *indirect* competition, and *direct* competition is extremely important in the development of my argument, since wild chimpanzees live by indirect competition, while post-1965 Gombe and post-1968 Mahale apes were forced into direct competition. It is hardly necessary to point out that indirect competition lacks the stress and tension of direct competition.

Charisma and dependence

Other key terms requiring definition are charisma and dependence. Max Weber (1957) originally defined charisma in a very narrow sense, as a state or quality produced in humans by the manifestation of some divine endowment. Sociologists now usually define charisma more broadly, as being the quality which is imputed to individuals, objects, actions, roles and so on, because of their presumed connection with fundamental, vital, order-determining powers (Shils 1965). Shils (1965:200) suggests that there is a widely dispersed 'normal' form of charisma present in the routine functioning of all human societies. This *normal charisma* implies no degree of sacredness, but a 'more attenuated, more mediated contact with such values or events through the functioning of established institutions.' According to Shils, this diffused 'normal' form of charisma is order-creating. It acts to maintain society, not to disrupt it.

Dependence, of course, is reliance on another for support of some kind. Because normal charisma is an attenuated form, connected with a role, we can expect that there are many chimpanzees in any group who possess charismatic attributes to varying degrees. I suggest that the dependence, too, of individuals taking that status/role is characteristically of 'normal' or routine attenuated form – an attribute of a role taken from time to time by all 'average' (i.e. not atypically nervous (or confident) adult chimpanzees and human foragers).

There is a problem with the concept of dependency, Bowlby (1973) points out. Too often, dependency has a negative connotation in our society. Rather than thinking of dependence as just relying on another for support of some kind, we tend to think of dependency as implying being subject to the relied-on person's jurisdiction, under their control.[2] Accordingly, we connect the state of being dependent with a subjective experience or expectation of weakness, helplessness and, often, insecurity. Bowlby (1973:37) feels that a pejorative view of dependence is so widely and uncritically accepted that it is necessary to substitute terms such as 'trust in' and 'reliance on' for 'dependence,' in order to be understood properly when writing on the topic. It is this very positive, normal form of dependence, i.e. trust in and reliance on others, without any connotation of weakness, that is characteristic of the mutualistic charismatic–dependent relationship.

The functional approach

A main interest of social anthropology is the classification and functional analysis of social structures and social systems. To the anthropologist, the traditional functional position is that societies should be viewed as systems having definite structure and organization, within which all major social patterns are interrelated and operate to maintain the integration or adaptation of the larger social system. The assumption is that social patterns can be explained by the effects of the pattern and that the consequences must be beneficial and necessary to the proper functioning of the system (Cancian 1968).

The concept of functionalism has been criticized and come into disuse because it is difficult to apply to complex human societies, in which conflict and change seem to be a normal condition. Another (strangely ethnocentric) criticism leading to functionalism's demise is that it justifies colonial rule, in that it provides scientific rationalization for maintenance of the status quo. Maintenance of the status quo is not always or by definition a bad thing. Indeed, functionalist concepts seem to be useful tools for analyzing the social behavior of undisturbed chimpanzees and human foragers who – under normal conditions – live successfully and peacefully by an uncomplicated, little changing, immediate-return foraging system.

For example, personality can be explained in functional terms. Social behaviors, relationships and mechanisms can be explained in terms of the adjustive and adaptive part that they play in an individual's total functioning. Because the social system of chimpanzees is based wholly on behaviors, interactions and relationships, these phenomena are both functional and structural elements of the adapted system.

Use of the term foragers

When referring to gathering–hunting peoples, the more general term foragers is used throughout in preference to gatherer–hunters because the nomadic gatherer–hunters have always subsisted very largely on gathered wild plant foods. Hunted meat has probably always been a prized occasional, rather than a staple, food (Lee 1968). Hence, the patterns of behavior of these gatherers are related to an omnivorous foraging mode of life, rather than a hunting mode.

As used here, the term foragers refers to human gathering–hunting groups that follow a nomadic pattern of fission and fusion, are without one permanent leader, and live on immediately obtained unprocessed wild food: that is, those who live by what Woodburn (1982) refers to as the 'immediate-return' system. While chimpanzees, too, live by the immediate-return system, when the single term 'foragers' is used in this work, it refers to these human groups.

The (sometimes) harmless people, and the (normally) friendly chimpanzee

When the Marshall family began their pioneering field studies of the !Kung San (1951–61), these people were living as foragers, in complete independence, wholly by gathering and hunting wild food (Marshall 1976:12). At that time, the arid land of the Kalahari Desert did not attract white settlers or native farmers and herders, and as Crown land the area was closed to white settlement. Since 1960, the first Commissioner of Bushman Affaires has been appointed, and the modern world and all its problems has intruded upon the once independent !Kung (Marshall 1976:14). Their lives have been vastly changed since the Marshalls visited them. Accordingly, the Marshall family's publications are based on the !Kung as they were, not as they are now. Early researchers write of gentle, friendly, harmless !Kung people, who abhorred and feared conflict and who placed a high value on peace and harmony (Thomas 1959; Marshall 1960, 1965, 1976; Silberbauer 1981, 1982).

Tanaka (1980:112–13), who studied the G/wi and the G//ana San groups for two 16 month periods in 1966–8 and 1971–2, reports that at the time there were almost no signs of competition or greed among the San, that rather than trying to outdo one another they strove to live in harmony and 'on the same level with others.' They avoided 'at all costs aggressive behavior toward their human colleagues' (Tanaka 1980:113).

But, as we know, by the 1970s, most of the San peoples had passed from 'a situation of local autonomy to one in which the direction of their lives came increasingly under outside control' (Lee 1979:420). Their traditional independent foraging way of life is now gone. Even in the 1960s, there was 'a pattern of aggression avoidance among hunters and gatherers in refuge areas but . . . a contrasting pattern of aggression (war, competitiveness, etc.) among hunters and gatherers in more homogeneous intercultural environments' (Gardner 1968:341).

Lee (1984:91) researched the period from 1922 to 1955 (inclusive) and learned of 22 cases of homicide among the !Kung during that 35 year period. The !Kung people now have a high rate of homicide. Now they are 'among the most violent people on earth. *In the space of a few short years their whole attitude toward violence has been transformed* and their homicide rate has tripled' (Lee 1984:150 (my emphasis)).

At the 1965 conference on 'Man the Hunter', Turnbull argued against the rising assumption that foraging peoples are normally aggressive. He pointed out that among the Mbuti pygmies 'there is an almost total lack of aggression, emotional or physical' (Turnbull 1968*b*:341). In an earlier publication, Turnbull (1961) makes it clear that it is managing to minimize outside influence and retaining control over their lives in the forest – rather than an actual lack of aggression – that permits the Mbuti still to live as 'harmless people.' As Gardner (1968) points out, Turnbull's nonaggressive Mbuti are in a refuge area, little disturbed by contact with outsiders. Perceptively, this scientist suggests that this difference from the coerced !Kung may be because these now aggressive, competitive !Kung groups are under intercultural pressure with which they cannot cope. Gardner's suggestion is in accord with the biological view. Biologists now believe that a society or group may be social or asocial according to environmental circumstances.

What of the reports from the nonprovisioning naturalistic field studies of 'the friendly chimpanzee,' which, because of the conflicting reports of the aggressive, directly competitive, artificially fed Gombe and Mahale apes, are assumed to be unrepresentative, hence of little scientific value?

Some readers may feel that the highly positive nature of the behavior of wild (not artificially fed) chimpanzees is overemphasized in this volume, that the 'picture' constructed from the naturalistic field studies of the friendly, nonviolent chimpanzee is unrealistically idyllic and one-sided. I think not. I do not suggest that the volatile, emotional chimpanzee is *by nature* a peaceful species, but they have the potential to be so. Indeed, the recent Gombe and Mahale evidence is that they also have the potential to be extremely violent and aggressive.

Perhaps only those who have studied the human foragers before their traditional way of life was disturbed – who are, as Turnbull (1983) suggests – familiar with the lengths to which humans, no better or worse than ourselves, subject to the same human temptations and weaknesses, will go to ensure nonviolence – will find it easy to accept the argument of almost totally nonviolent, only indirectly competitive, chimpanzee groups. Only further naturalistic field studies will settle the question.

Still, I do not view the apes and the humans living by the highly positive egalitarian mutual dependence system as idyllic societies, in the sentimental 'noble savage' tradition. *But, the mutual dependence system is extremely efficient, under normal circumstances.* The argument here is that, in normal socioecological conditions the adapted social system works to defuse and diffuse dissent before it builds up, explodes into aggression and destroys the harmony which is necessary to the maintenance of the very loosely structured social group. Positive behavior is a seminal property of the social structure.

In the least complex of the small-scale human societies, those who live by what Woodburn (1982) refers to as the immediate-return foraging system – which is outlined in Part 2 – the emphasis is on interpersonal and intergroup relations, to which the social system is subordinated. Turnbull (1972) points out that both security and survival is seen in terms of these relationships, with the result that the foragers tend to display positive characteristics that we find admirable: kindness, generosity, affection, hospitality, cooperation, egalitarianism and others. He suggests that this 'sounds like a formidable list of virtues, and so it would be if they *were* virtues, but for the hunter (forager) they are not. For the hunter in his tiny, close-knit society, these are necessities for survival' (Turnbull 1972:31). Accordingly, Turnbull (1968*a*) suggests that our understanding of immediate-return foraging organization will be enhanced if we think of such attributes as being *essential properties* of the social structure, not just as values and norms; because, lacking such (structural and functional) social attributes, the fluid, loose social system would collapse.

Human foragers are normally free to exit from a situation which even hints at tension, and they usually do so, Turnbull (1972) submits. Although we cannot know their motivation, chimpanzees too constantly enter and leave groups. It seems that various easily activated – and easily overlooked – forms of voluntary exit (fission) are a fundamental device. The result (among the few who have studied wild chimpanzees) is an impression, which is somewhat misleading, of these animals living, in Ghiglieri's (1979:236) words, 'in frictionless peace.' In fact, there is a constant redressing of the delicate balance of the egalitarian society among both species of immediate-return forager. In this book the nonaggressive wild groups which were studied by naturalistic methods are not idealized, but presented as they were observed, functioning smoothly in their adapted mode. Further field studies of undisturbed wild chimpanzees will prove – or disprove – this assertion.

While we must be aware of the biological contribution when considering primate behavior, when discussing the social behavior and organization of chimpanzees we must assign both social and ecological conditions as much weight as we assign to these phenomena among humans who follow a basically similar foraging adaptation. The portrayal of human foragers as 'harmless people' and of chimpanzees as aggressive, dominance seeking, territorial and occasional eaters of their own young, are both true and false depending on the circumstances.

The naturalistic studies: practitioners and methods

Carpenter (1965:256) holds that in the ideal naturalistic field study, 'the observer should make his observations of a primate in its native habitat without disturbing or modifying the behavior.' However, since this is almost impossible when the subject is a wary, human-shy wild primate species, he suggests that the practical aim of naturalistic methods should be to reduce observer disturbance to the lowest possible degree.

The principal observers using various naturalistic methods whose work is referred to in this study include: Jane Goodall (prior to 1965) and many Japanese observers, some of whom carried out several field studies, both separately and in partnership, prior to beginning regular artificial feeding in 1968. These include S. Azuma and A. Toyoshima, Junichero Itani, M. Kawabi, Akira Suzuki and Toshisada Nishida. (For dates, partnerships, locations and durations of chimpanzee field studies up to the end of 1972, see Baldwin and Teleki (1973:315–30)). The naturalistic studies are often criticized as being short term and fragmentary, but all of the above observers (with the exception of Kawabi) carried out at least one naturalistic field study that was, by Baldwin and Teleki's (1973) criterion, long term; that is, lasting for periods of from 12 to 24 or more months.

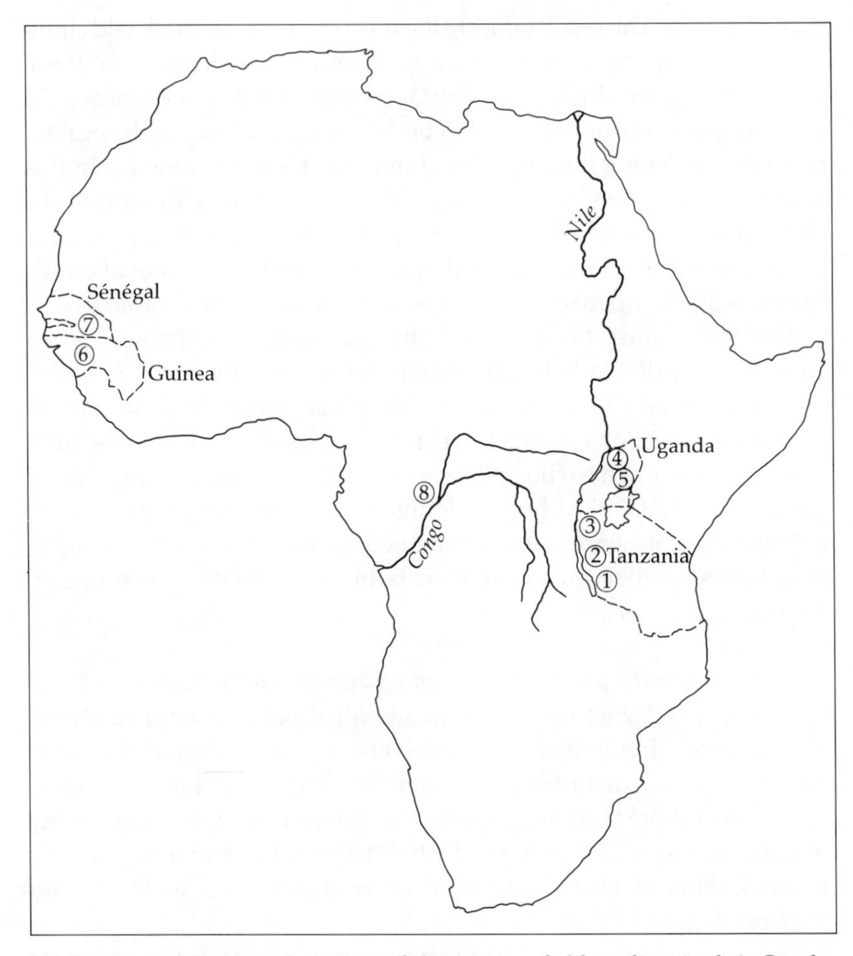

Figure 3. Approximate locations of chimpanzee field studies cited. 1, Gombe Research Center; 2, Mahale Research Center; 3, Ghiglieri; 4, Sugiyama; 5, Reynolds and Reynolds; 6, Albrecht and Dunnett, Kortlandt; 7, Baldwin; 8, Kortlandt.

Other principal naturalistic studies relied on include those by Vernon and Frances Reynolds, Adriaan Kortlandt, Yukimaru Sugiyama, Y. Sugiyama and J. Koman, H. Albrecht and S. C. Dunnett, P. J. Baldwin, and M. P. Ghiglieri. The last was a long-term 18 month study. All studies were temporally and geographically separate, some by up to 5000 kilometers (see Figure 3).

Despite these entirely separate studies being uneven, and carried out under differing conditions, literally all of the naturalistic researchers report that interactions which appeared to fit the dominance concept formed but a minute fraction of the observed (wild) chimpanzee behavior. All report the association pattern of the group that they studied to be one of very temporary, flexible, ever-fissioning and fusing small subgroups within larger, bisexual, open local groups of highly associative yet independent individuals. With the exception of Nishida, no researcher using naturalistic methods reports any sign of a dominance hierarchy, closed group, one alpha leader or attempt to maintain exclusive sexual access to receptive females. (It was Nishida's (1970) impression in the early years that, although the apes were rarely truly aggressive and there was no permanent leader, some extremely vague indications of a dominance hierarchy were discernible.) All report little or no aggressive interaction (Azuma and Toyoshima 1962; Kortlandt 1962; Reynolds 1963; Reynolds and Reynolds 1965; Goodall 1965*a*; Kawabe 1966; Itani and Suzuki 1967; Nishida 1968, 1970; Sugiyama 1968, 1969, 1972; Izawa 1970; Albrecht and Dunnett 1971; Kano 1971; Baldwin 1979; Ghiglieri 1979). Although they all report the same positive social behavior – which is not at all like the recent aggressive, dominance-oriented, fiercely territorial behavior that the Gombe and Mahale observers record – the data from the naturalistic studies are summarily dismissed as yielding 'little or no data' on behavior, sociality or ranging patterns (Ghiglieri 1984:2).[3] The truth, however, is that, the neglected naturalistic field study publications are rich with evidence which supports a mutual dependence argument.

Field studies of chimpanzees of longer duration than Nissen's (1931) 64 days of observation began in 1960 with Kortlandt's in the eastern Congo (Zaîre) and Goodall's at the Gombe Stream Reserve (now Gombe National Park), Tanzania, and Itani's in the vicinity of Mahale Mountains, also in Tanzania.

Kortlandt spent a number of months traveling throughout the chimpanzees' natural African range, looking for an open area regularly visited by the apes, from which to conduct his observations. He finally found it on an abandoned section of a pawpaw and banana plantation which edged the forest in the eastern Congo. By habit, wild chimpanzees came regularly from the forest to feed on the fruit; and the plantation owner did not interfere with their doing so.

Kortlandt built several small hut-like blinds and observation posts, some in trees, 20 to 80 feet (6–24 m) off the ground. In his highest blind, 80 feet off the ground, he overlooked a large area in all directions. Once in his chosen post, Kortlandt usually remained concealed until nightfall so that the chimpanzees would not see him and perhaps become alarmed and not return to the area.

Unlike Kortlandt, who made no attempt to follow the apes in their forest wanderings, the team of Vernon and Frances Reynolds attempted a more difficult task for 8 months in 1962. Guided by the animals' vocalization, they followed the often fast-moving apes through the dense underbrush of Budongo Forest in northwestern Uganda. Concealed in the foliage, the Reynoldses observed the wary chimpanzees when they stopped to feed, rest or socialize. This proved to be so difficult, the apes often slipping away just as the weary but determined Reynoldses arrived, that they also employed native watchers whom they stationed in areas through which the chimpanzees were known to pass habitually. Neither of these techniques appears seriously to have disturbed the wild apes.

Yukimaru Sugiyama has carried out long-term naturalistic studies of the behavior of wild chimpanzees in a number of locations including Bossou, Guinea, and Budongo Forest, Uganda. He also intruded as little as possible on the wild natural groups. For his 7 month study in Budongo Forest in 1966–7, he chose a chimpanzee group that was somewhat accustomed to the presence of people, in that there were several native villages close to the animals' natural range. Without feeding the apes, Sugiyama was able to follow and observe them from fairly close quarters, for long periods, without unduly alarming them.

For 6 months in 1968–9, Helmut Albrecht and Sinclair Dunnett studied the behavior of two separate groups of chimpanzees in the Republic of Guinea, at the most westerly limit of the apes' natural range, 5000 kilometers from the Gombe and Mahale centers. Like Kortlandt, they chose a deserted portion of a (grapefruit) plantation which edged the forest, where the local chimpanzees were accustomed to come to feed. They cleared a rough field for better observation and encouraged the apes to feed there by gathering and piling mounds of grapefruit within view of the blinds and concealed cameras. Although this brought the animals to an area convenient to the observers, they were not frustrating the apes, since there was no human control over the supply, no limited ration, and the apes had earlier incorporated a visit to feed on grapefruit at the location into their regular foraging rounds. The animals did not lose their autonomy over the food supply. Any ape who preferred to was free to pick grapefruit elsewhere, other than from the observers' heaps. Albrecht and Dunnett (1971) noticed that some of the chimpanzees picked up and ate the fruit at the heap, but most of them gathered an armful and ate it at the edge of, or in, the forest. The apes were usually selective in choosing the fruit, touching several before selecting – a leisurely pursuit which certainly does not suggest tension.

Despite the fact that they emphasize in their book a general similarity between the observed behavior of the Gombe apes and their study group, Albrecht and Dunnett (1971:118) report that there were 'a few differences' also. Among the differences cited are: (1) the Guinean chimpanzees are apparently not hunters, opportunists or organized; (2) there was no clear hierarchy of dominance; and (3) the animals' sexual behavior was clearly different from that reported at Gombe. Goodall (1968*a*) reports that some Gombe females when in estrus were seldom able to move away from a spot without being followed by the males. Albrecht and Dunnett (1971:118–19) observed no such behavior in either of the two separate wild populations they studied. 'Oestrus ♀♀ seemed to come and go in the same loose way as other adolescent and mature individuals.' While at Gombe males took the initiative in 83% of the copulations or attempted copulations which Goodall (1968*b*) observed, Albrecht and Dunnett's (1971:119) data from observation of one group, 'while insufficient, indicate that only about one-quarter of the copulations observed there were initiated by the ♂♂'.

Albrecht and Dunnett feel that their study was limited by its restricted length – 3 months in each of two Guinean locations totalling 6 months in all – and suggest that a much longer study would be required to establish 'if such differences as were observed are real or merely apparent' (Albrecht and Dunnett 1971:10). Nevertheless, these attributes and behaviors – the relaxed sexual and nonhierarchical behavior and relationships and general lack of tension of the Guinean chimpanzees – are strongly similar to those reported by virtually all of the naturalistic observers of undisturbed wild chimpanzees.

Other naturalistic studies referred to in developing this study include that of Pamela Baldwin, whose 16 month study in 1976–8 as a member of the University of Stirling Primate Research team was of an isolated group of wild chimpanzees in a dry mixed habitat of woodland, strips of gallery forest and open vegetation near Mount Assirik, Sénégal. Baldwin's group was largely undisturbed, since they inhabit a remote area where no humans live and few come. Because of the difficulties of attempting close observation of the extremely shy, widely roaming apes without artificially feeding them, Baldwin's study is a natural history of this wild group. Her method of observation was the same ideal naturalistic method attempted earlier by the Reynoldses' team. She, too, followed the chimpanzees as they moved about, trying to remain out of the animals' view, hiding behind natural features, usually observing from a distance through binoculars.

Michael Ghiglieri's is a socioecological study. His interests were behavior, sociality, grooming, ranging patterns and the social system. Ghiglieri spent 18 months (between 1976 and 1978) observing 'truly undisturbed, unprovisioned chimpanzees,' whose home range lay in the center of the nature reserve in Kibale Forest, Uganda (Ghiglieri 1979:277). There, in order to study monkeys, earlier researchers had laid out a rectilinear grid of trails covering an area of about 6 square kilometers. Ghiglieri used this grid, which connected a number of favorite feeding sources of natural foods used by both monkeys and chimpanzees, as his study area. He accustomed the apes to his presence by regularly and openly appearing at these favored feeding places. He reports that he seldom attempted to follow the chimpanzees when they left this grid-area (Ghiglieri 1979, 1984). For this reason, he emphasizes that his data on ranging patterns and territorial behavior of chimpanzees are, 'equivocal', 'weak' and too few (Ghiglieri 1979:259, 271).

Although they sometimes supplied bait food to the wild chimpanzees, early studies by Japanese scientists *prior to 1968* – such as those by Azuma and Toyoshima (1962), Kawabe (1966), Itani and Suzuki (1967) and Nishida (1968, 1970) – are, in the terms used in this volume, also naturalistic studies. Any artificial feeding was short term, not very successful (not accepted by the apes), and access to the bait was not withheld by the humans, so the feeding in itself seems unlikely to have had a significant effect on the behavior of the chimpanzees.

By far the longest, most continuous detailed data on the behavior and organization of undisturbed wild chimpanzees are Jane Goodall's publications based on observation of the Gombe apes from 1961 to 1965, before the frustrating 'closed-box' method of feeding was introduced. As well as writing numerous scholarly articles on her Gombe studies, Goodall wrote a fascinating account of her experiences with the Gombe chimpanzees for the general reader. This book, *In the Shadow of Man* (Goodall 1971*a*), was developed in terms of the distinctive personalities of the individual, now well-known apes.

Like field researchers following her, Goodall found that ideal naturalistic methods – following and observing shy wild apes on their daily rounds, while in no way interfering with them – was extremely difficult and expensive in terms of time and research funds. She sought to habituate the apes to tolerate her presence without provisioning, but found habituation a lengthy process. She testifies that the first 5 or 6 months of her studies formed a period of 'initial fear and hasty retreat' on the part of the wild chimpanzees (Goodall 1971*a*:68). Fortunately, she found a hill peak above a grove of fruiting trees which attracted the animals daily while the fruit was ripe. Sitting quietly at a considerable distance, observing their interactions through binoculars, Goodall became familiar and less threatening to the cautious animals.

It took 14 months of persistent following before the chimpanzees would carry on their normal activities (feeding, mating, sleeping, and so on) while Goodall sat quietly 30 to 50 feet (9–15 meters) away. Although no longer frightened by her stationary presence, the wild apes still showed alarm if she attempted to follow them (Goodall 1965*a*).

It was a yet further 8 months before the calm, confident mature male she dubbed David Greybeard began to visit Goodall's camp, initially to feed from a fruiting tree that grew inside the area. On one of his visits he took a banana from Goodall's table, and thereafter she made sure that there were bananas available whenever this confident ape appeared. For the next 5 months this animal came to camp alone. Then, David began arriving at the camp accompanied by another adult male, quickly dubbed Goliath because of his superb physique. Still later a third, very timid, adult male, whom Goodall called William, joined them. For some time the latter two apes were nervous; they watched from the safety of the trees. Eventually, they joined the bolder David, feeding on the banana 'bait' within the bounds of the camp. For over a year these three were the only chimpanzees to venture into Goodall's camp.

In 1963, a confident elderly female, sometimes referred to by Goodall (1971*a*) as 'fearless Flo', and her three offspring began to visit the camp along with the three males. When Flo went into a period of estrus that same year, a 'whole retinue' of males, females and juveniles followed her to camp (Goodall 1967*b*:43). At that point there were 20 chimpanzees of both sexes and various ages visiting the camp. In publications based on data collected at Gombe before 1965 (when the observers took control of immediate access to the desired bait food away from the apes), the behavior Goodall reports is strongly similar to that reported by other observers using naturalistic (nonfeeding) methods.

Feeding methods: the Gombe Research Center

From August 1962 to March 1965, Goodall fed any of the chimpanzees who came to her camp, simply by placing in several locations ample, open heaps of bananas from which the animals foraged in quasi-natural fashion. The chimpanzees who came to feed sat around the piles of fruit, taking from them freely, without direct competition or possessiveness (Goodall 1963*a*). This method, designated System A (Wrangham 1974), was the first of five provisioning systems tried at Gombe between 1962 and 1965.

However, even 20 chimpanzees is a much larger number than was usually seen together by Goodall prior to 1962, and by the end of 1964 there were 45 recognized chimpanzees regularly visiting the feeding area within her camp. Unfortunately, as Goodall soon found, one adult chimpanzee could – and did – consume 60 or more of the fruit at one sitting. So too could a baboon, a large troop of which Goodall found she was unwillingly provisioning along with the chimpanzees. And the bananas had to be imported and paid for.

Since such feeding put a strain on the limited research funds, in 1965 a new system (B) was introduced. For this system a number of concrete feeding boxes were made and sunk into the ground. Thin steel lids were held shut by underground wires attached to handles some distance away, enabling the enclosed banana supply to be doled out by the humans, at moments deemed to be suitable. The researchers usually sprung the lids open at irregular intervals when not too many baboons were around.

By controlling access to the bananas the observers hoped to lessen the aggressive competition over the limited bait that had sprung up between the apes and the baboons, and to ensure that every chimpanzee got the contents of one box each day. But this hope was frustrated by the direct competition of all of the apes over each box that was opened (Wrangham 1974:84). Goodall (1971*a*:143) realized that:

> the constant feeding was having a marked effect on the behaviour of the chimps. They were beginning to move about in large groups more often than they had ever done in the old days. They were sleeping near camp and arriving in noisy hordes early in the morning. Worst of all, the adult males were becoming increasingly aggressive. When we first offered the chimps bananas the males seldom fought over their food; they shared boxes . . . .

[Now] not only was there a great deal more fighting than ever before, but many of the chimps were hanging around camp for hours and hours every day.

The nomadic fission and fusion pattern entailing constant change of companions and disbursement and movement about the home range was already being disrupted. The aggressive direct competition 'eventually created so many problems that observation was almost ended' (J. Goodall, personal communication, cited by Wrangham 1974:85).

This system was in force when the first of a wave of independent researchers arrived in June 1967. By that date, 58 chimpanzees of all ages were regular visitors, some coming almost daily (Goodall 1986*b*). Until then, Goodall, photographer Hugo van Lawick, two research assistants (after 1966, apparently several assistants), a secretary and a camp staff of local people were the only workers at the Gombe site. By June 1967 the Gombe apes had been experiencing the aggression and competition-rousing method of provisioning for more than two years (Baldwin and Teleki 1973:319; Goodall 1971*a*:131). Consequently, *all studies carried out at Gombe, other than Jane Goodall's pre-1965 work,*[4] *are of chimpanzees that had already experienced prolonged, human-imposed interference with access to a desired food.* By 1967, the interactions and relationships of the Gombe apes were very different from those reported by Goodall in the 4 years prior to 1965, and it was in 1967 that the systematic, much relied on, data bank was begun.

In August 1967, feeding system C was introduced. From August 1967 to June 1968, bananas were provided in the boxes for a day or two; then the boxes were left empty for a few days, in unpredictable sequence (Wrangham 1974). The objective of System C was to reduce the time that the apes spent at the feeding station and, in this way, reduce the aggression. However, the introduction of the 'No P' (no provisions) days did not reduce the amount of aggressive competition on the days that the boxes were filled. In fact, on the provisioned days the local baboons were now competing even more boldly and aggressively with the chimpanzees, Wrangham (1974) reports.

When Goodall returned to Gombe in 1968 after a term of study in England, she found that the feeding situation was in 'chaos' (Goodall 1971*a*:145). She reports that the observers tried to discourage the baboons by refusing to open laden boxes while they were near camp, 'but this merely built up tremendous tensions and frustrations in the chimpanzees who were there. Baboons and chimps alike knew that there were bananas in the closed boxes, and the longer we delayed opening the worse the situation became' (Goodall 1971*a*:146). When the boxes were finally opened, there was 'bedlam'.

System D was tried from July 1968 to June 1969. It entailed feeding selected aggregations of the apes with one box per individual, approximately every 10 days. The chimpanzees could smell the bananas in the boxes, but could not obtain them. This prompted the chimpanzees who smelled the contents to wait around, hitting and prying at the boxes with sticks (Wrangham 1974:85). For the food-autonomous apes, this was a new and utterly unnatural situation.

System E was introduced in June 1969, and was still in effect when Wrangham left in 1972. The human-controlled feeding was continued, this time using boxes embedded in the walls of a covered trench. The observers had access to the back of the boxes while remaining out of sight, and so they could fill a suitable number of boxes when a group of apes arrived and they judged it expedient to feed them. They tried to provide food so that each ape was artificially fed but once a week. The observers found the goal of this system impossible to attain.

The aggressive direct competition which the incoming observers (mainly students) observed among the Gombe apes fitted well with the dominance theory then taught; hence, there seemed little reason for the newcomers to question the normality of the behavior of the provisioned primates. To repeat, because it is a vitally important fact that has been virtually overlooked; no researcher, other than Goodall (from 1961 to 1964) has studied the Gombe apes in the 'wild' state. Literally all other Gombe studies, including Goodall's post-1965 to the present-day studies, are categorically provisioning studies of disturbed chimpanzees.

Feeding methods: the Mahale Research Center

While the field studies by Japanese scientists *prior to 1968* are categorically naturalistic, all post-1968 Mahale studies are provisioning studies. The methods used to feed the Mahale chimpanzees bear little resemblance to methods used at Gombe, yet they too resulted in frustration and aggressive competition among the local groups.

The Mahale studies also began in 1961. Between that year and 1965, a number of Japanese scientists carried out short-term field studies of chimpanzees in the general area where the planned permanent Mahale Field Research Center was later established (see Baldwin and

Teleki 1973). Until 1965, naturalistic methods were used, the observers attempting the difficult task of following the wild apes, without artificially feeding them (Nishida 1979). But this method produced only very brief opportunities for observation. Even after the animals were successfully habituated to the presence of the observers, it was found that attempting observation without the lure of bait foods produced only very short, fragmentary observations of behavior. Accordingly, part of the plan for the permanent Mahale Research Center was to seek to habituate the chimpanzees through supplying bait food at a fixed feeding area, in a manner that would not disrupt their natural foraging ways. To this end, in October 1965, about three acres (1.2 ha) of underbrush was cleared, and seedling and mature sugar cane plants were planted in the clearing, with the aim of providing a desirable permanent crop which the chimpanzees could forage in fairly natural fashion. As a regular supplement, several hundred ripe sugar cane stalks were scattered in the forest beside the planted field (Nishida 1979). Within 6 months, some chimpanzees had begun to venture into the cleared area to feed on the sugar cane plants. To facilitate closer observation, a further 50 stalks of cane per day were scattered on a regular basis, in a narrow area within 80 meters (later 10 to 30 meters) of an observation scaffold. Observation at this fixed feeding station was the main study method in the earlier years after the permanent center was established.

Like bananas, sugar cane does not grow wild in the area, so unfortunately for what is a commendable plan, the chimpanzees quickly destroyed the plants by overfeeding. Cut cane continued to be supplied, but, in 1968, it became very difficult to obtain, so the amount of bait put out had to be reduced. Nishida (1979:77) reports that direct competition for the now insufficient supply of sugar cane sticks caused a high frequency of aggressive behavior, and that sometimes 'subdominant or shyer' apes were unable to obtain any of the desired bait. In a study in which he sought to establish the psychobiological bases of aggression, Moyer (1976:190) indicates that 'when a goal presumably attractive to two or more subjects is made available in such a way that it can be obtained by only one of them, the end result is frustration for the loser.'

Because of the increased aggression, when some of the Mahale apes became habituated to the observers, a mobile provisioning method was begun, in which the observers carried a small supply (one to five cut-up stalks) of sugar cane into the forest, and mimicked the chimpanzees' hooting call. The habituated animals learned to come to the observers, seeking the cane (Nishida 1979).

'Heavy' artificial feeding at a feeding station was discontinued after 1968, but was reintroduced for one particular study of the Mahale K group in 1972–3 (Mori 1982:46). Mori reports that at the completion of this 6 month study the feeding station was abandoned. However, in studying the other group (M), 'intensive artificial feeding' was carried out in the periods 1968–70 and 1972–4, 'by providing sugarcane and bananas at a fixed feeding station' (Hiraiwa-Hasegawa, Hasegawa and Nishida 1984:402). Since 1975, the extent of feeding has been reduced and observations are made mainly in the forest, Hiraiwa-Hasegawa *et al.* (1984) assert. But the frustrating method of withholding part of the baits and supplying a small ration from time to time was continued as part of the forest feeding, according to Nishida's (1980) report.

As part of the post-1968 provisioning methods, both at the Mahale feeding station and in the forest, the observers withhold the food from the apes until they think all have arrived, and the initial excitement of the animals subsides. At the feeding station this is a period, usually, of about 10 minutes (Mori 1982). Then the first cut-up cane (of a ration of four or five at the station) is scattered. The usual method in both locations is to give the assembled animals a few sizable pieces of the bait, then to suspend further provisioning for a time during which the observers give their full attention to recording the chimpanzees' behavior. At the feeding station, the withheld stalks are given out at a rate of about one ration per hour (Mori 1982).

These delaying tactics keep the apes in the vicinity for several hours, facilitating lengthy, continuous, concentrated observation. Similarly, in the forest, also, some baits are withheld. This causes the chimpanzees to 'become frustrated because there are many baits visible in front of them' (Nishida 1980:124). They engage in frustration behaviors and displays, interpreted by Nishida as 'food-demanding', since they are directed toward the observers and enacted only when the scientists are withholding a sizable portion of food. The very different feeding methods at the Gombe and Mahale centers share this element of blocking the motivated apes from their goals of obtaining the bait foods. Psychologists have established that a situation in which a motivated individual is blocked from achieving a sought-after goal contains the constituent elements for producing frustration. In fact, both Goodall (1968*b*) and Nishida (1980) cite this inability of the apes to reach the food that they desired and knew was there as examples of a frustrating situation for the animals. The theory regarding behavioral changes brought about as a response to frustration will be discussed in Part 4.

Obtaining food is considered to be the most basic of physiological needs. Interference with the ability to do so can be expected to be more deeply disturbing than overcrowding, or even restricting access to mates. In the wild, food may be scarce from time to time, but, as we know, the natural form of food competition of chimpanzees is *indirect*, the separate simultaneous seeking of the same resources (Mayr 1970). This situation – being aware that a desired food is present but not obtainable except at unpredictable intervals and through direct competition – is abnormal. Wild chimpanzees do not, in nature, experience this loss of autonomy in regard to a food supply. Other than in experimental circumstances, even captive monkeys and apes are seldom tantalized by the bafflement and frustration of present and unobtainable food, irregularly made suddenly available. Deliberate irregularity in feeding time forces the animals to remain tense and vigilant throughout the whole indeterminate waiting period, increasing the stress level of the whole experience (Averill 1973).

Assignment of watershed dates (Gombe, 1965; Mahale, 1968) at which the behavior of the Gombe and Mahale apes ceased being essentially normal, adaptive behavior and became increasingly behavioral responses to frustration, is of course arbitrary and hypothetical. The change was gradual. Goodall herself places the date when she began to see changes in many aspects of the behavior of the Gombe apes at 1964. In a 1979 publication she stipulates that prior to 1965, from '1961 to at least 1964', the males ranged farther than they are known to do today (Goodall *et al.* 1979:21).

Another change 'since 1964' that Goodall (1968*a*:292) reports is in the nature of baboon/chimpanzee interaction. Prior to that date chimpanzee 'children' and young baboons occasionally interacted playfully. The most frequent play patterns Goodall observed consisted of chasing, leaping from branch to branch, with quick touches. 'Since establishment of the feeding area', Goodall (1968*a*) writes, wrestling, tickling and play-biting have been observed 'more and more frequently during such interspecific play sessions.' Play between the young seemed to be becoming focused increasingly on directly competitive 'play-fighting' rather than on testing their own limits and those of others. 'Prior to the setting up of the artificial feeding area I saw few aggressive incidents' between chimpanzees and baboons, Goodall

(1968*a*:291) reports, but since then aggressive incidents have increased both in frequency and violence.

My argument regarding extensive, unrecognized (negative) social change as a result of frustration is centered largely on the Gombe evidence. One reason for this emphasis is that the closed-box feeding methods begun at Gombe in 1965 are model situations for inducing frustration and aggression. Another reason is that certain continuing behavioral differences between the two provisioned groups – continued inclusion of the Mahale females as members of a bisexual group, for example (Nishida 1979), suggests that the Mahale feeding methods are perhaps, or so far, less disturbing to the social order.

Some researchers, uneasy with the possible effects of provisioning, have warned that artificial feeding may have affected and changed certain aspects of the animals' usual behavior. Pusey (1979) suggests that this human interference may have retarded the normal social development of youthful apes. Others suggest that it may have caused the phenomena of female exogamy and male retention (Sugiyama and Koman 1979), obscured assessment of the normal position of the female in the community (Riss and Busse 1977), or broadly affected the social and ecological patterns of the community (Reynolds 1975). These warnings have not been taken seriously. The majority of Gombe and Mahale researchers assume that, while provisioning does frustrate the apes and causes some changes in their behavior, any changes manifested by the apes are quantitative, not qualitative, and that the behavior of these animals accurately reflects the 'intrinsic qualities' of the species (Nishida 1979:79).

This assumption is understandable in terms of the times and the (then) state of our knowledge. When field studies of primates became suddenly very numerous in the early 1960s, the concept of social or hierarchical dominance was still an enormous influence on scientific thinking. The dominance concept was developed in 1922 by Schjelderup-Ebbe, who worked on the pecking order of chickens, and extended by Zuckerman (1932) to primates, in his study of the organization and behavior of a well-fed but exceedingly crowded baboon aggregate in London Zoo. By the time field studies of chimpanzees began in the 1960s, the concept of social dominance had become a trusted 'paradigm' – Kuhn's (1970) term for the conceptual tools which most scientists in a discipline use at a particular time.

Kuhn (1970) has pointed out that a strong commitment to the accepted conceptual or instrumental framework of a paradigm on the part of scientists is a necessary part of the scientific method. Normal or ongoing science cannot proceed without such framework and commitment. Still, Kuhn warns that the theoretical, instrumental and methodological commitment has the power to blind adherents. It usually results in the accepted paradigm becoming the filter through which practitioners approach scientific problems, to the point where their perception may be affected adversely. This leads to an 'immense restriction of the scientist's vision and to a considerable resistance to paradigm change' (Kuhn 1970:64). Perhaps similar overcommitment to the concept of social dominance is among the reasons that an influential segment of the academic community continues to rely on the recent Gombe and Mahale findings for their understanding of chimpanzee behavior and social organization, to brush aside the warnings, and to ignore the contradictory findings from naturalistic studies of wild chimpanzees.

Accordingly, a main task is to show, in as much detail as possible, the extent and cumulative effects of the social change brought about by the frustrating Gombe and Mahale feeding methods. To make the differences in behavior, and social and emotive atmosphere of the wild and the provisioned groups quite clear, in areas where comparable published evidence exists, the comparative method is used. Data on particular aspects of social behavior from the naturalistic and the provisioning studies are placed in sequence, for comparative purposes. Evidence is marshalled which shows that an almost all-encompassing negative social change spread in ripple fashion to affect vital aspects of social behavior and organization of the Gombe and Mahale groups. It is argued that consequently the social behavior and organization of these sample populations are not representative of that of undisturbed wild chimpanzees. The other principal task is to develop a model of the adapted form of social organization of wild chimpanzees (and undisturbed human foragers) from the evidence found in the publications based on naturalistic field studies. This shared fundamental adaptation is referred to herein as mutual dependence. Before these tasks are taken up, anthropological understandings regarding the most simple full form of human foraging organization are outlined.

PART 2

The human foragers

General anthropological understandings

This part of the book is a brief summary of anthropological understandings regarding the social behavior and organization of foragers, gatherer–hunters, who lived by the most simple full form of human organization, the immediate-return system, at the time they were studied. More detailed data from the field studies of these groups are inserted throughout the text, wherever they illuminate aspects of the social behavior and organization of chimpanzees.

Anthropologists consider the ancient foraging (gathering–hunting) way of life as the most successful and longitudinal adaptation humans have ever achieved. Indeed, until the advent of the age of agriculture, approximately 10,000 years ago, all peoples everywhere on earth lived by a foraging mode (Murdock 1968). Today, foraging societies have all but vanished. There remain only a few, small, scattered groups, in environments climatically so inhospitable as not to be much desired by agriculturists or industrialists, and even these areas are being pre-empted rapidly (see figure 4).

There is a general agreement among anthropologists that all foraging peoples live in small groups, and follow a nomadic pattern known as fission and fusion. Another point on which there was, until recently, complete agreement is that, whatever the foraging construct is, it 'is not a corporation of persons who are bound together by the necessity of maintaining property' (Lee and DeVore 1968:8).

There is considerable diversity among foraging models. Some anthropologists suggest lineage models, local groups which are organized around central members of a single lineage, and some who affiliate through marriage alliances. Others propose territorial models,

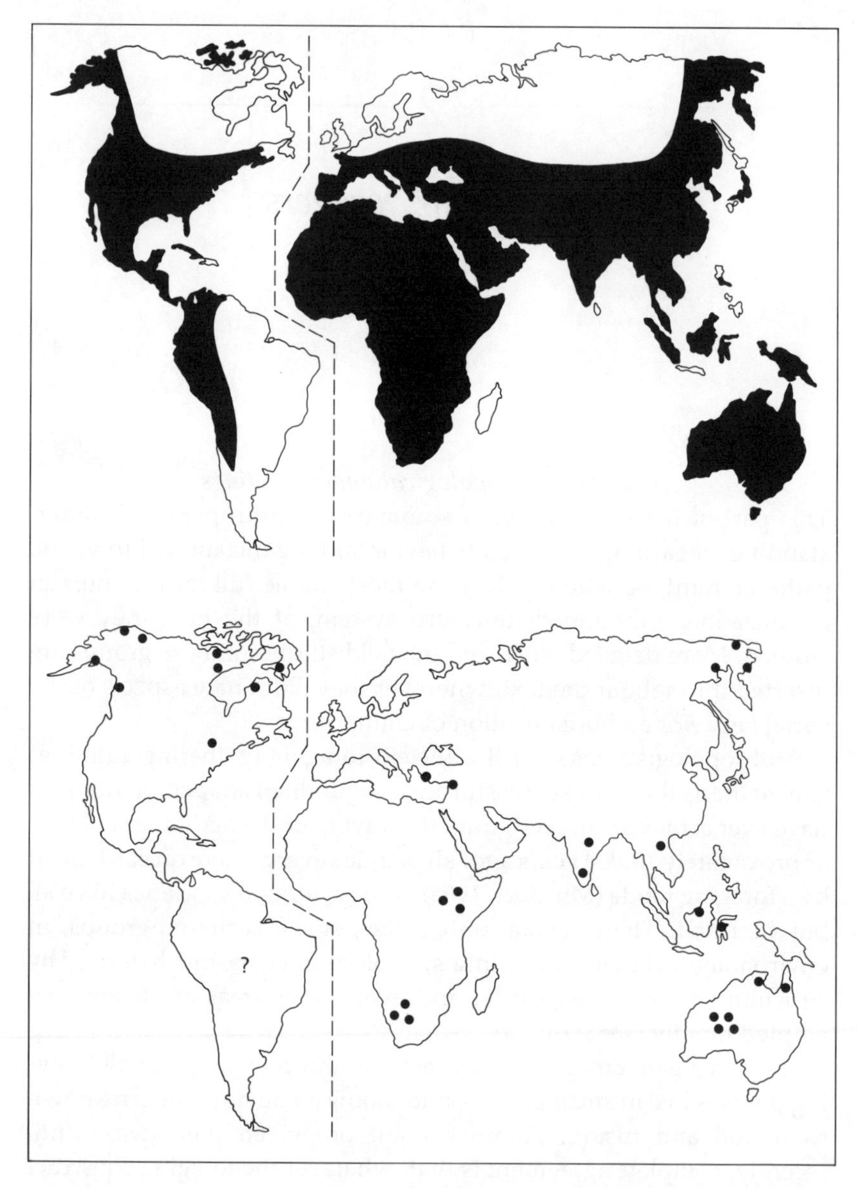

Figure 4. The decline of human foraging (gathering–hunting) societies. Top: 10 000 BC; world population 10 million, 100% foragers. Bottom: AD 1972; world population 3 billion, 0.001% foragers. (Adapted from Lee and DeVore 1968: Frontispiece).

agricultural and fairly sedentary groups with fixed camp locations. The development of the argument in this study would not be advanced by exploring the full range of possibilities. Only the least complex form, Woodburn's (1982) immediate-return system of foraging organization, is pertinent.

Reynolds (1966) and Itani and Suzuki (1967) were probably the first to point out that wild chimpanzees and some foraging peoples have many similar organizational principles and behavior patterns. Reynolds suggests that human foraging society evolved out of nomadism, open groups, lack of territoriality, loosely defined home ranges, a wide recognition of relationships, and the inheritance of behavior patterns such as tool use, drumming, dancing and bedmaking of an ape-like ancestral stock. Undisturbed chimpanzees manifest all of these phenomena.

As indicated in Part 1, this study extends the similarities much further. It is argued that both six remaining foraging peoples, and undisturbed chimpanzees, live by what Woodburn (1982) refers to as the immediate-return system. Hence, the immediate-return system will be outlined at this point.

The immediate-return foraging system

Woodburn (1982) divided human foraging societies into two broad general categories – immediate-return (which is not an association of persons joined to maintain property) and delayed-return systems (in which some property is maintained). In groups that live by various delayed-return systems, people hold rights over valued assets such as beehives, pit traps, fish weirs and so on, which yield a delayed return for considerable labor over time. Delayed-return systems have certain basic implications for social relationships and social groupings, Woodburn suggests. They imply binding commitments and dependencies between people and a set of ordered, jurally defined relationships through which goods and services are transmitted. It is in the small, more frequently studied, delayed-return societies combining agriculture and herding with hunting and gathering that 'we find the familiar kinship commitments and dependencies; lineages, clans, . . . marriages in which women are bestowed in marriage by men to other men' and so on (Woodburn 1982:433).

Woodburn (1982) points out that immediate-return systems do not entail binding commitments, that the members could not incorporate them and retain this independent way of life. Because our interest is in the immediate-return system, no material on foraging people living by any of the variety of delayed-return systems is used in this study, and they need not be discussed further.

By the 1980s there remained only six widely separated small-scale human groups who live by the immediate-return foraging system. Woodburn (1968*a*,*b*; 1970, 1972) lists these and the scientists who have studied them: the Mbuti Pygmies of Zaire (Turnbull 1965, 1966); the !Kung San of Botswana and Namibia in the Kalahari Desert (Marshall 1976; Lee and DeVore 1976; Lee 1979); the Hadza of Tanzania (Woodburn 1968*a*,*b*, 1970, 1972); and the Batek Negritos of Malaysia (Endicott 1979). Woodburn (1982) also draws on Wiessner's studies of the !Kung people in 1977, K. L. Endicott's work on the Batek Negritos in 1974 and 1979, and Morris's study in 1975 and Gardner's in 1980 of the Pandarm and Paliyam of south India. As inferred, wherever human foragers are referred to in this study, reference is to only these small groups who still follow the immediate-return system of foraging. Any data used will be from studies of some of the above groups, or be general understandings that pertain to the immediate-return foragers' way of life.

Although the surviving human immediate-return foraging societies are geographically widely separated, their societies are organized around the same few simple principles, which Woodburn outlines. According to Woodburn (1982:432), groups following the immediate-return system have the following basic characteristics.

1 Members go hunting or gathering and eat the food obtained the same day, or casually over the next few days. Thus they obtain an immediate and direct return from their labor.
2 Food is not elaborately processed, nor is it stored.
3 Tools are simple, portable, utilitarian, of local materials, hence easily acquired and easily replaced. Like weapons, they are made with skill, but without involving very much labor.

Wild chimpanzees meet all of the above criteria for being categorically an immediate-return foraging society.

In the immediate-return societies no formal systems of law exist. There are no specialized or formalized institutions or spheres that can be designated as political, economic, religious, judicial and so on. Hence the social sphere must be capable of dealing with all problems which, in a more complex society, would come under the jurisdiction of one of these institutions. Accordingly, such foraging societies are organized on a basis of social relationships and required behavior: roles and statuses, enforced by a few simple but effective positive and negative sanctions (rewards and punishments). To repeat (because it cannot be overemphasized), certain behaviors, interactions and relationships are not merely valued or desirable, but *necessary*, for the society to function. Put another way, the social structure serves as an end in itself. *The totality of necessarily positive social relations between all members of the immediate-return societies constitutes the social structure.* 'Formal government would destroy the egalitarian nature of the society, and, lacking egalitarianism, the cooperative effort would collapse' (Turnbull 1968*a*:24).

As anthropologists use the term, an egalitarian society is one in which all members are considered to be of equal intrinsic worth and are entitled to equal access to, and share of, the goods, rights and privileges of their society. Within the structure of the society, individuals have a high degree of autonomy. The fundamental immediate-return foraging system eliminates power relationships, and inequalities of possessions, prestige and status. Indeed, Woodburn (1982:432) emphasizes very strongly that no other way of human life 'permits so great an emphasis on equality.'

Woodburn (1982:434) finds the 'profound egalitarianism' of the surviving immediate-return foraging peoples deeply impressive. Equalities (of power, rank, prestige and wealth) are not just taken for granted, but constantly reasserted. As Service (1966:83) explains, egalitarianism in foraging bands is expressed through individuals being 'strongly resistant to direction through authoritarian power of any kind.' All adults are autonomous (self-direction is another property of the system) and highly 'individuated.' They are not equal in the sense of everybody being similar to one another physically and psychologically, nor in everyone possessing the same status or amount or type of influence. But because of the uninstitutionalized, unspecialized nature of the society all adults participate much more fully in every aspect of the culture than do the people of more complicated societies. Service (1966) emphasizes that individuals are what they are, skilled in some things and less skilled in others; and they are fully accepted on that basis, in their natural variety.

'Systems of immediate-return offer the individual, so long as he retains his health and strength, a rather special type of personal autonomy and security', Woodburn (1980:106) reports. People's lack of dependence on others for access to resources, and the ease with which an individual can segregate his or her self from anyone with whom he or she is in dispute without sacrificing important interests, greatly reduce the probability of conflict (Woodburn 1980:106). Another result of the expectation of high individual independence is that each adult is self-directed and responsible for his or her self. The Hadza carry this idea of self-responsibility to the point of abandonment of the seriously ill, even when they are close kin (Woodburn 1980:105).

Fried (1967) defines the egalitarian society as being one in which every person has a position of prestige. He suggests that 'an egalitarian society is characterized by the adjustment of the number of valued statuses to the number of persons with the abilities to fill them' (Fried 1967:33). Like Service, Fried suggests that some members are less assertive or less skillful than others; but, he emphasizes, egalitarian societies have powerful leveling mechanisms that prevent any emphasis on gaps in ability among members. Because there is no means of fixing or limiting the number of persons exerting influence or authority through knowledge, skill, personality and so on, he suggests that a dominance order cannot appear in conjunction with such a system. There is no superior–inferior ordering based on physical dominance or other sources of power such as wealth, hereditary classes, military or political office in these foraging societies (Service 1966:31).

In immediate-return societies, the basic, essential egalitarianism is also promoted through the patterns of mobility and flexibility of groups and companions (fission and fusion), and of environmental equality – free, direct access to food and other resources (Woodburn 1982:434). It is shown in this volume that essentially the same form of egalitarianism is promoted among chimpanzees, through the same patterns and the same expected behaviors and sanctions.

Human social organization at the most simple, fundamental level is, of course, much more complex than that of the chimpanzees. Human foragers have marriage, nuclear family units, a powerful ethic of organized reciprocal sharing of meat, ceremonies, dances, ritual healers and the greater communication of language. The humans have a division of labor, and chimpanzees do not. Traditionally, men hunt and women gather, but most anthropologists now agree that while this division is clear, it is pragmatic, rather than rigid. In the literature, much emphasis has been on the (male) hunters, but for the most part gathering and hunting are jointly cooperative activities and neither carries greater prestige (Turnbull 1968*a*:23).

Men may occasionally gather (although not with a women's group) or they may remain in camp, in preference to hunting. Although women are opportunist hunters of small game, they do not join men's hunting groups. However, the pattern of movement and structure is not different. Both men's hunting and women's gathering groups are *mobile* units, while those who choose to remain in the temporary camp – the sick, the elderly, the child tenders, who may include any of the men who are not out hunting – comprise a *sedentary* group. I suggest that this division by activity, rather than by male and female sex as such, 'fits' better with the concept of a profoundly egalitarian society. 'Both women and men participate in decision-making and conflict resolution and are influential in accordance with their age, experience and wisdom' (Leacock and Lee 1982:10).

The nuclear family is the primary human economic and traveling unit, but single families or small groups of two or three families may form a subgroup to forage together for a time. The larger society congregates only periodically, in seasons and areas where food supply is ample. These gatherings (camps) typically remain for a matter of days, weeks or even a few months, although there is a constant change of personnel through people flowing in and out.

Territoriality

Participants in discussion at the 1965 symposium on 'Man the Hunter' agreed that human foraging groups do not function as closed social systems. They suggest that probably right from the very beginning there was communication between groups, including reciprocal visiting and marriage alliances, so that the basic foraging society consisted of a series of local 'bands' which were part of a complex network of interconnected and interbreeding, linguistic communities (Lee and DeVore 1968; Turnbull 1972).

From the forementioned determined basis, Lee and DeVore (1968) visualized a generalized model of the basic human foraging system with the following characteristics. To begin, the necessity of moving around in order to get wild foods would keep personal property at a very low level. This would keep wealth differences from developing, and help the society to maintain a basic egalitarianism, as well as facilitating freedom of movement.

The scattered nature of the food supply would keep groups small, usually fewer than 50 persons, although groups would wax and wane in size. Smaller groups would come together at intervals forming a larger social group. Variations in regional wild foods would be the basis for the fluid situation. Obtaining a variety of foods would be best handled by small subgroups moving about the land, from one area to another. Consequently, local groups would not control access to resources, and there would be a lack of concern and of possessiveness in regard to food resources. Mutual access to resources of all kinds would be highly important as a means of communication between groups because, when exclusive control of resources and facilities begins, the loose non-corporate nature of the small-scale foraging society cannot be maintained.

Turnbull's (1968*a*) summary of general anthropological understandings regarding the social organization of contemporary human foraging groups is in agreement with the findings of the 1965 symposium on 'Man the Hunter', but he extends a few points. He suggests that the constant process of fission and fusion (of both local groups and the larger society) goes beyond mere response to locating and obtaining food, a view which Woodburn (1982) shares. Human foragers (and chimpanzees) tend to move long before the local food supply is exhausted, and as suggested, a pattern of constant realignment of individuals and groups acts to nullify conflict, and the development of any corporate, unilateral ranked system such as a dominance hierarchy. Social flexibility allows for optimal exploitation of the preferred effective resources by the total population, including the ill and the elderly (Lee 1968, 1979).

Human foragers following their nomadic life do not wander aimlessly across the face of the land. They have a home range: familiar areas which they customarily use. There is no clear border to such ranges, nor to the resident population. Both are vague, usually overlapping entities, without very definite boundaries. Groups are open, there are no exclusive qualifications for membership provided immigrants conform to the expected behavior (Silberbauer 1981; Woodburn 1982). Territory is determined to a large extent by natural barriers, such as hills, rivers, and ravines, and also by the migratory habits of the game (Turnbull 1968*a*:24). Turnbull makes the pattern of residency very clear when he explains that a territory or home range may be fixed – that is understood, but the population it contains is not fixed (Turnbull 1972:299). People flow back and forth freely and spontaneously. The composition of a population is entirely unpredictable. 'In many instances, it is difficult to decide just what constitutes a "population" in either biological or sociological terms', Turnbull (1972:292) indicates. Chimpanzee organization is similar in this respect. Nishida (1979:115) points out that 'unit groups' (local groups) are permanent in the sense that each 'is connected closely with a given territory and lasts permanently and independently of individuals, even if members of the group change constantly.'

When Lee began his studies of the then isolated population of the !Kung San in 1963, they were still living their traditional foraging way. Lee (1979:457) reports that a 'dynamic of movement coupled with the face that both males and females form the core of groups leads to an emphasis in social relations on recruitment rather than exclusion' in the egalitarian society. He remarks that in contrast to the old model of male-centered territorial bands proposed by Service (1962) and Tiger (1969), 'the maintenance of flexibility to adapt to changing ecological circumstances is far more important in . . . [forager] groups structure than is the maintenance of exclusive right to land.'

Social kinship

Kinship in human groups is not merely or always biological. It is frequently cultural or social. Clans, totems, tribes – categories of people considered to be kin – vary independently of actual consanguineal relationship. Many societies organize around such social kin categories.

The immediate-return human foraging groups are organized on a basis of kinship relations; but virtually everyone is considered to be kin, and treated as such, regardless of actual biological relationship (Marshall 1957, 1960; Turnbull 1972; Lee 1979; Silberbauer 1981). Despite the prominence and cohesiveness of the nuclear family economic unit, kinship in these societies is basically social rather than consanguine. Among the Hadza, such social kinship unites groups scattered, until recently, over more than 1000 square miles (259,000 hectares) in Northern Tanzania. The !Kung San local bands have a cohesive bond of social kinship with other !Kung groups living as far distant as 80 to 115 miles (128–184 kilometers), through a special name-relationship. They do not consider encountered persons from other groups as strangers, but 'of their own people,' even if they have not previously met (Marshall 1957). It is not that kinship is unimportant, Woodburn (1980:105) points out, 'only that its significance is different.' It is quite usual for kinship terms to be widely used to address and to refer to other members of the community; 'indeed, many of these systems are universal kinship systems in which everybody . . . is able to define a kinship or quasi-kinship tie to everybody else' (Woodburn 1980:105).

Woodburn (1980:105) makes the nature of the kin relationship even clearer when he explains that the term kinship is a metaphor for *connectedness* when used in reference to foraging peoples. Such kinship provides 'a set of rough and ready expectations for appropriate behavior,' (Woodburn 1979:245), a 'broad idiom for friendly rather than hostile relations' (Woodburn 1980:105). It gives the individual 'a sense of identity but provides him (or her) with no substantial rights,' and 'no substantial duties' (Woodburn 1980:97). The extended kin relationship is one of nonantagonistic mutuality, without any strong commitment to specific individuals. There is an emphasis on group, more than consanguineous, ties. Social relations that cut across biological and family ties are peace-making devices, Lee (1968) points out. They permit people to enter and leave a group with a minimal disruption of the social order.

Turnbull (1965) suggests that a more elaborate kinship system would be a disadvantage because the composition of the social group changes constantly. He also implies (Turnbull 1968*a*:23) that while the cooperative nature of the foraging economy, and the egalitarian sharing of food may be responses to the diurnal economic system (daily food-searching), both mitigate 'against the formation of exclusive groups, kinships or otherwise.' A structured lineage system involves specific loyalties. This would 'fragment the group into opposed sections that would have no structural validity,' Turnbull (1965:291) explains. If this holds among human (immediate-return) foragers, it casts doubt on the current assumption of the normality and importance of consanguineous kin relations among wild chimpanzees, who also live by the immediate-return foraging system.

Leadership

The immediate-return foraging group is a consensual polity (Turnbull 1968*a*; Silberbauer 1981). Nowhere in these societies do we find a secular authority backed by power (Turnbull 1968*a*). There is no permanent leader. Indeed, the constant change of leaders gives the appearance of there being none, Woodburn (1982) suggests. Moreover, in the immediate-return human societies, leaders are 'very elaborately constricted to prevent them from exercising authority' – other than some individuals being deferred to by others, and having their opinions carry a little more weight than the opinions of other discussants (Woodburn 1982:444).

Anthropologists usually refer to such leadership as charismatic, a matter of having influence over others but no direct authority or power to coerce. It was suggested earlier that this is charisma of the 'normal,' diffused, order-creating form that Shils (1965) suggests is found in all human societies. Such charismatic leadership is based on personal attributes, personality, skills, knowledge, wisdom and so on. It is temporary and shifting according to circumstantial needs.

Silberbauer's (1981) description of the nature of leadership among the G/wi !Kung in the period of his field study (in 1958–66) before 'radical change' had occurred in the traditional pattern of !Kung life is particularly illuminating. He reports that leadership was diffused as well as temporary (Silberbauer 1981:462). It was 'an ephemeral role' adopted by one or another member 'in response to a particular situation when his or her opinion or advice is adopted by the rest of the band.' Individuals, men and women, sometimes attract others because they are skilled hunters (and their advice is valued, as when hunting parties are forming) or renowned healers, or exceptional story-tellers and so on, rather than being solely attractive on a basis of personality traits. Consciously or unconsciously following Darwin (1871), Service (1966:31–2) points out that such attributes do not confer leadership 'unless these qualities are put to work in the service of the group.' The leader role is not sought after, and those who are in the role change constantly, as needs and circumstances change. The leadership role is spontaneously assigned by the group, conferred on some member or some members in some particular situation, not taken (seized) by an individual. Because a person's opinion on the best hunting strategy or cure for an illness is accepted and acted on does not mean that his or her views on other matters will carry weight. One leader replaces another as needed.

Leadership is persuasive, not authoritarian, and 'serves only to guide the band toward the consensus that is the real locus of decision' (Silberbauer 1981:462). Therefore, others' 'autonomy and self-reliance are not diminished by dependence on the authority of . . . [any] central or prestige figure' (Silberbauer 1981:462.) There is no loss or gain of self esteem through taking either leader or follower role. The emphasis, in these societies, is on generalized mutuality, rather than on specific, individual commitment (Woodburn 1980).

In functioning immediate-return foraging societies, enculturation is for avoidance of conflict and competition. This attitude is implicit in the behavior, interactions and relationships the young see around them. Among these humans, this attitude is not consciously taught; it comes about through observation and imitation (Draper 1978).

Because aggressive actions are avoided and devalued, and negatively sanctioned by adults, the children have little opportunity to observe or imitate overly aggressive behavior. In the traditional way of life, the children of foragers grow up in an intimate, peace-maintaining, care-taking, egalitarian social environment. The young absorb the positive, nonaggressive, noncompetitive attitudes of the adults who surround them (see Montagu 1978).

It will be shown in Part 3 that this is the wild chimpanzee child's experience also. Both young wild chimpanzees and children of human foragers have the same attitudes and behaviors to learn: attitudes and expected behaviors that permit both societies to develop their potential for harmonious relations between themselves and their neighbors. As we shall see, also in Part 3, the socialization experiences of the post-1965 Gombe young are opposite in tone and effect.

The foragers' sanctions

In the egalitarian foraging societies all adults have high autonomy. There is nothing other than the social sphere – roles, statuses and relationships – regulating any aspect of the social organization. There are no viable structural alternatives. Egalitarianism requires equality in goods and resources as well as in status and rank, hence the essential controls are entirely social in nature. They are (1) positive sanctions – rewards or 'reinforcers' to use Baldwin and Baldwin's (1981) term; and (2) negative sanctions, punishment, dissuaders. These are essentially the gaining, expressed through both individual and group consensual behavior, of acceptance and approval (reward), or the denial thereof (punishment). Among humans, the latter is expressed through shunning, shaming, mocking, group ostracism, driving out and in very rare, extreme cases, execution through group action. Both positive and negative sanctions are extremely effective in these societies, and can be expected to affect the self-esteem of the sanctioned individual.

The prime reason behind all sanctions, both positive and negative, Turnbull 1968*a*:24) asserts, 'is the maintenance of the delicate equilibrium that enables a hunting and gathering band to pursue its essentially cooperative, egalitarian economy.' In normal conditions, i.e. the uncrowded natural world to which they adapted, foraging societies are in delicate balance, in the sense of fine rather than fragile. The balance or equilibrium of these societies is easily disturbed, but they incorporate social mechanisms that normally keep the domain of stability broad and flexible enough to respond to, and absorb, whatever social disturbances and geophysical crises may occur. Despite their apparent lack of structure, these small loosely organized societies are exceptionally strongly bonded. Turnbull (1968*a*:25) suggests that 'their very survival is sufficient testimony to the effectiveness of their informal structure and the directness of their response to the environment in which they have to live'.

Turnbull (1983:28) points out that, while we humans seem to have a limitless capacity for violence, for aggression, the small foraging societies show that we have an 'equally great potential for nonviolence and nonaggressivity.' A conspicuous feature of the 'small-scale' (foraging) societies is the great amount of concern shown for the reduction of the human 'violent potential' for aggression to a 'remarkable minimum' (Turnbull 1983:28–9).

It is not that these foraging peoples were or are any more moral than ourselves, Turnbull (1983) continues, 'if they see the wisdom of minimizing violence and aggressivity . . . to a level far below their mental and technological potential, it is perhaps simply *because that best answers their overall needs for survival*' (my emphasis).

PART 3

The changing social order

In this section of the book the behavior and social organization of wild chimpanzees, as reported by those using naturalistic methods of field study, are compared with the same phenomena as they are reported among the Gombe and Mahale chimpanzees. The reader is warned that the way the evidence of negative social change is gathered together and presented here gives a very one-sided picture of the Gombe situation. Positive (friendly, supportive) interactions have not ceased. (They may even have increased as a result of the need for greater reassurance and consolation). There was not unremitting turmoil, but sporadic eruptions, with intervals of peaceful behavior. This comparison is begun by reference to an early insight of Kortlandt, as it structures the adapted egalitarian system of chimpanzees.

Kortlandt's hypothesis

Kortlandt's startling hypothesis is based on his 1960 observations of wild chimpanzee groups, made largely when they came to feed on pawpaw fruit growing on an abandoned edge of a plantation at Bossou, New Guinea. It was to his own very great surprise, Kortlandt (1962:132) comments, that he realized that 'the chief social distinction' in wild chimpanzee groups 'is between childless and childrearing adults rather than between males and females.' While this is an enormously important insight, it was met with a resounding silence by the academic community, presumably because this vital understanding does not fit within the then-unchallenged dominance paradigm.

Most of the early observers categorized chimpanzee groups into four types of subgroup, usually mixed groups (males, females and young) mother groups (females and young), all male, and lone male groups (Goodall 1965*a*; Reynolds and Reynolds 1965; Sugiyama 1972; Ghiglieri 1984). But Kortlandt (1962) suggests that there are only two types of subgroup. (He reasons that lone individuals are not a type of group, and that a mixed group of both sexes and all ages represents the society, not a type of group).

The two types of group that Kortlandt (1962:132) identifies within the society are, first, a 'sexual group' – 'mainly adult males and childless females' (not necessarily in estrus) which also occasionally included a few females with young; and, second, a 'nursery group,' consisting of juveniles and their mothers, 'and sometimes a few adult males,' – essentially *childless* and *childrearing* groups. The division into these two subgroups came about only when the apes began to move, and participation in either group did not seem to be fixed or controlled in any way, Kortlandt submits. Groups that were exclusively male, or male with one or two estrous females attached, were not seen by either Kortlandt, or Albrecht and Dunnett (1971), who studied the same group in 1968–9.

Kortlandt's division into (childless) sexual and (childrearing) nursery groups is made on the basis of the two groups behaving differently, and because these aggregations were apart more often than they were together. Kortlandt does not specify how the behavior of these groups differs, but one obvious way is in the degree of mobility. The nonchildrearing adults are obviously more able to be mobile, to range more swiftly and widely than childrearing females who, during the frequent, lengthy periods in which they are engaged in nurturing dependent young, will tend to be more sedentary.

Both Nishida (1979) and Ghiglieri (1984) note that childless adult and youthful nulliparous females usually range widely with the adult males, rather than less widely with the mothering females. They 'become regular members of the core-subgroup' (Nishida 1979:97). Rather than being all-male, the wide ranging, 'core-subgroups' are 'male-estrous female' in composition, Nishida (1979:107) maintains. Goodall (1965*a*) identified similar Gombe groups in the early years as being male-adolescent rather than male-estrous female in composition. Youthful chimpanzees of both sexes often moved about with mature males, 'the adolescent female [doing so] whether or not she had a sexual swelling' (Goodall 1965*a*:461). While often the youthful females in these wide-ranging groups will be in a state of estrus, I suggest that the female's being in a state of youth, in transition from childhood to adult rank, is more significant in structural terms than her being in estrus. While females carrying an infant, or matching their pace to a walking, still-dependent child, do usually travel less extensively 'than do males and females without infants,' Goodall (1965*a*:455) points out that 'this depends in part upon the individual animal.' A few occasionally choose to travel widely, with a swift-moving mobile group.

While childless females traveling with the largely male mobile group are in a minority, they are not deviant, or exceptional, in this behavior. Through observation and by 'aunting', youthful females do learn the female parental role by remaining with the more sedentary mothers; whereas youthful males must separate from their mothers and join the mobile males in order to find their role models – and to learn the geography of the home range and variety and location of food sources. But if the youthful female remained solely with the sedentary group throughout her nulliparous period she would lack this vital botanical and geographical knowledge. It is reasonable to assume that this would make the female apes, as a sex, dependent on the males' food-finding. This dependence would undermine the adaptive self-sufficiency of adult chimpanzees in food-finding.

Accordingly, it seems that a period of travel as part of the actively roaming groups would be typical, not exceptional, behavior of youthful nulliparous females. I again suggest that it is not puberty as such, but social maturity which is reached at about the same time as puberty, that inclines young females to join the mobile groups, and the wild adult males to accept them as they accept youthful males of similar age and appropriate social maturity.

As adolescent (youthful) females have much to learn about mothering from the childrearing females, the (many) mobile groups are unlikely to contain all of the youthful females of the group at any one time. A large numerical majority gives the active groups the appearance of being male groups, in that they contain most of the males from adolescent or apprentice age to fairly old, plus the (exceedingly few) sterile or childless adults, and youthful females who choose to travel with the mobile members to learn 'what's out there' – something those taking the mobile role can teach them.[5]

Obviously, the mobile (childless adult) group will, in most instances, be composed of a very large majority of near-adult and mature males, and a minority – perhaps only one or two – female chimpanzees. However, the youthful females who join this type of temporary traveling group are not auxiliary or peripheral. Their status, and that of the youthful males, might be likened to that of (human) apprentices. This argument is expanded later (see Part 3, Apprentice into adolescent).

Reynolds and Reynolds (1965) estimate the total population of wild chimpanzees in their study area as being 70 to 80 animals. Of these, 20 or more were adult males, while only 6 to 10 were adult or near-adult females without young. Thus, if all of the childless apes of both sexes were to travel together in one group – and they never do – the group would consist of at least twice as many males as females. In relation to their total numbers, the 'few childless females' who are members of a mobile adult group might represent a greater percentage of such females, than do the males, of the total male group membership.

There are social and ecological reasons (as well as reproductive opportunities) which help to explain the presence of quite young females in mobile groups. This is discussed in Part 6. Accordingly, I suggest that a modal mobile group would consist of any adult males and childless females (sterile, temporarily childless, or near-adult in age and not yet pregnant) who are physically vigorous, able and temperamentally inclined to travel rapidly and widely, for a time.

The current, much-subscribed-to understanding is that the active, widely ranging subgroups are all-male groups. Recently, Ghiglieri asserts that, as a result of recent Gombe and Mahale and Kibale studies, the structure of chimpanzee communities is coming into sharper focus. A consensus is emerging that 'genetic relatedness and inclusive fitness are key factors in the evolution of a [chimpanzee] community maintained by males' who cooperate to exclude others from their territory, its resources, and opportunities to mate with females of the group (Ghiglieri 1985:111–12). (It will be argued in Part 4, that such exclusive behavior is not a good, or even adequate, strategy to forage their particular type of habitat).

The division proposed by Kortlandt (1962), between the adult chimpanzees engaged in child rearing and those which are not – hence sedentary and mobile groups – seems more in accord with the high egalitarianism and lack of exclusion characteristic of immediate-return foraging groups than does a division into male and female groups; particularly since the apes do not have the male hunter/female gatherer, and nuclear family, divisions of the human foragers. In the next section the structure of chimpanzee subgroups shall be reconsidered in those terms – as mobile and sedentary, rather than as male and female, groups.

Mobile and sedentary groups

In the early years, Nishida (1968), like Goodall, emphasizes that wild chimpanzee subgroups are not organized on the basis of age, sex, blood relations or sexual attraction, but seem to be entirely randomly composed. He reports Mahale subgroups consisting of five males and one adolescent female, six males with two females with infants and so on (Nishida 1968:215). When Itani and Suzuki (1967) studied wild chimpanzees, before provisioning began, they encountered both what appeared to be all-male groups and mixed groups. On one occasion, when a mixed group was passing them on the far side of a ravine, the males dropped out and remained watching the observers as the other apes passed by. It is Itani and Suzuki's (1967:364) impression that it was encountering them 'that made the adult males congregate and assume a defensive attitude,' and that conceivably a congregation solely of males may form as a result of anticipated or actual threat or danger to the group. This, too, is an important and neglected observation.

Both Kortlandt (1962) and Reynolds and Reynolds (1965) report that the mother or nursery (sedentary) groups often contain a few adult males. Nishida (1968) also reports a few females in the 'adult' groups, but because these groups are, in numerical terms, male, he suggests that such adult groups 'may be regarded' as being a male group (Nishida 1968:185).

Without doubt, an observer might frequently see groups consisting of only one sex, all-female or all-male, and the ranging, swift-moving groups are, as we know, largely male (see Table 1). Because there are *numerically* more males, these latter adult groups have gained wide acceptance as being, in terms of *structure*, male groups. The concept of single-sex groups (male-only and mother groups) as characteristic of chimpanzee social organization has been assumed on the basis of numerical predominance of one sex in such groups. However, when Reynolds (1963:97) postulates an all-female (mother) group, he also notes that 'very old males, and timid males, seem to attach themselves to groups containing mothers and other females,' not acting as leaders, 'but just as companions, as though they preferred the peaceful life' (Reynolds 1968:211). Reynolds (1968:211) observes that a group of two or three mothers with their offspring might remain in the vicinity of a

Table 1. *Composition of wild chimpanzee groups*

References	No. of groups observed	Groups of 6 or fewer[a] (%)	Mixed groups (all ages, both sexes) (%)	Sedentary Groups ('Mothers and dependent young') (%)	Mobile groups (Adult males and adolescents of both sexes) (%)	Males only (%)
Goodall 1965*a*:452[b]	350	82	30	24	18	10
Nishida 1968:182–3[b]	218	55	52	14	11[c] included as male only	
Reynolds & Reynolds 1965:398–9	215	62	41	18	24[d]	16
Izawa 1970:24	35	NOT GIVEN	43	6	26[e]	3
Suzuki 1975:275	NOT GIVEN	NOT GIVEN	78	3	5[f]	1

[a] Data on size of parties (six or less) supplied by Nishida and Hiraiwa-Hasegawa (1986:167). For simplicity, all percentages are rounded off to the nearest figure.
[b] Gombe and Mahale sources used are based on data from pre-provisioning studies. Lone chimpanzees are not included in this table.
[c] Nishida lists under 'males only' some groups which include a few adolescent females or mothers with infants, but are 'numerically male' (Nishida 1968:185). (I suggest that these are mobile groups, as are all of the following.)
[d] 'Adults of both sexes, and occasionally adolescents, but not including any mothers with dependent young' (Reynolds and Reynolds 1965:398).
[e] Listed as 'adults only, both sexes' (Izawa 1970).
[f] Suzuki lists his data source as Suzuki (1969). His 'adults' category is assumed to be structurally males and adolescents of both sexes.

fruiting tree for several days 'while the *more energetic and mobile members* of the community, *the adult males, the childless females and the adolescents* would visit several different feeding places in the course of a day' (my emphasis). Elderly animals of both sexes, confident or timid, tend to range less widely than the younger (childless) apes (Kortlandt 1962; Goodall *et al.* 1979).

Reynolds' observation is important, and of more categorical importance than is generally realized. In effect, he too has categorized the chimpanzees into childless and childrearing groups, although he too, in keeping with the times, proposes the mobile active groups as 'bands of males' – often 'accompanied by' childless females (Reynolds 1965*a*:181).

Division by sex has blurred our understanding of chimpanzee social organization. The numerically male active subgroups are not structurally male, but those able and willing to move about rapidly over long distances, i.e. the mobile – the childless adults and near-adults of both sexes. Similarly the numerically predominantly female 'mothers' groups are structurally the more sedentary members – mothering females, the elderly of both sexes and any prime males who might choose to remain with a sedentary group for a time.

Sedentary and *mobile* are accurate categories in terms of function and structure. The important organizational roles connected with both of these types of groups are discussed in Part 6. There is, under normal circumstances, no social role among chimpanzees for all-male or all-female groups as such.

It was mentioned earlier that human foragers have a division of labor, by sex: men hunt and women gather. Yet, among these humans too, the fundamental division is by mobility, not by sex or task. Although men and women do not forage together, both women's gathering and men's hunting groups are *mobile* groups; while persons, of either sex, who choose to remain in camp form the *sedentary* group. These include resting hunters, and gatherers, some mothers, women in advanced pregnancy, the very young and anyone, male or female, incapacitated by illness, injury or advanced age.

The social climate of wild groups

Mischel (cited by Staub 1984:35) points out that psychologists sometimes classify human environments by their perceived 'social climate', dependent on 'the nature and intensity of personal relations;' and by such functional properties as the 'reinforcement consequences for particular behaviors in that situation'.

The 'social climate' of wild chimpanzee groups, i.e. the prevailing emotive tone, expressed through various social behaviors which distinguish and give a distinctive quality or ethos to the social system, is highly positive. Ghiglieri's (1979:194, 235) subjective impression is that the wild chimpanzees of Kibale Forest, Uganda, are 'happy-go-lucky' animals that live in a social atmosphere of 'frictionless peace,' usually free of even the most subtle sign of dominance and hierarchy. Ghiglieri's assessment echoes the reports of earlier naturalistic observers in other areas. In reference to one ape's usual 'extroverted, unconcealed joy' in being with preferred companions, Ghiglieri (1988:69) suggests that this animal's response is 'excellent testimony to the blessings versus the conflicts inherent to the ape's fusion–fission social system.'

Many of those who have studied the Gombe and Mahale groups find positive reports such as Ghiglieri's hard to believe, in view of the polar behavior and social climate of the provisioned groups at both research centers, in recent years. It is generally assumed that, if the naturalistic observers had had the benefit of the long periods of observation that provisioning facilitates, they would have seen very different behavior and organization.

Probably all free chimpanzees have been, to varying degrees, affected by human pressure. When Goodall began her studies in 1960, there were large tracts of undisturbed forest to the east of her study area, but Gombe is now a small national park of about 20 square miles (5180 hectares), surrounded by human habitation. Yet, despite occasional failures of main food sources, spatial erosion of natural habitat by human settlement, and the consequent moving closer of formerly more distant wild chimpanzee groups, there is (or was, until the mid-1970s) still ample wild food in the Gombe Park (Goodall 1971*a*; Riss and Busse 1977; Teleki 1975, 1981). Even during a 2 year drought in 1961–2, during which fruits did not form on the trees (and bananas were not yet provided), the Gombe chimpanzees stayed well-fleshed and healthy (Goodall 1965a). They did not compete directly for food, but simply changed their pattern of meeting periodically to feed in large noisy congenial groups to one of ranging silently through the forest alone or in very small parties of two to six, feeding more eclectically – a survival pattern of the human foraging peoples in similar circumstances. Goodall (1990:124–5) describs the supply of wild food as usually 'more than adequate for the requirements of both chimpanzees and baboons.' It is 'sometimes in relatively short supply' during the dry season. It seems likely that the coping pattern described above would still operate.

Because the observers do not allude to shortages of wild food in the quite similar Mahale habitat, I assume sufficiency in that area also.

In most areas in which wild chimpanzees have been studied, large troops of baboons share their habitat and food supply without aggressive competition for food or space (Goodall 1965*a*; Reynolds 1965*b*; Albrecht and Dunnett 1971; Ghiglieri 1979; Ransom 1981). Ghiglieri (1979) reports not only a high mutual tolerance but even a certain affinity of the two species. In one instance, he was convinced that two males, one a baboon and the other a chimpanzee, had been traveling throughout the forest together as comrades (Ghiglieri 1979:172).

In the pre-1965 years the chimpanzees and baboons at Gombe either tolerated each other's presence or interacted amiably. Even from 1967 to 1969 the two species spent by far the greatest amount of their time together either in essentially 'neutral' interaction (as feeding side by side) or in positive interactions, such as play between juveniles and even occasionally between adults (Teleki *et al.* 1976; Ransom 1981). After 1965 at Gombe, however, aggressive interactions between chimpanzees and baboons waiting around camp for the banana bait 'increased enormously in frequency' (Goodall 1967*b*:38). (Goodall (1971*a*) reports that the problem of competition for bananas between the apes and the baboons at the feeding station grew steadily worse every year). She comments that on her return to Gombe as early as 1966, after several months in England, she was 'horrified' at the change in the behavior of the animals.

None of the earlier field researchers who used naturalistic methods saw evidence of predation on the chimpanzees by any other species, even the human. At the time that the Reynoldses and Sugiyama carried out their studies of the chimpanzees in Budongo Forest, and of Kortlandt's, and Albrecht and Dunnett's Guinean studies, the apes were protected by colonial law and also by local superstition from human hunters. Local people neither considered them a desirable food nor took much interest in them (Reynolds 1964; Reynolds and Reynolds 1965; Kortlandt 1962; Sugiyama 1968; Albrecht and Dunnett 1971). Today they are hunted (Ghiglieri 1979), and crowded by human settlement.

In the forests the wild apes show no panic reactions to snakes, buffalo, elephant or to the only carnivore in the area large enough to be a potential threat – the leopard (Goodall 1965*a*; Nishida 1968; Ghiglieri 1979). While some chimpanzees give vent to alarm calls when they see leopards, others do not react or respond to the calls, but continue to feed, groom and so on (Goodall 1968*a*; Izawa and Itani 1966).

Goodall (1965*a*) and Reynolds and Reynolds (1965) remark on the chimpanzees' surprising lack of alertness in the Gombe and Budongo forests, and suggest this as evidence that the wild apes are not afraid of being attacked. Wild chimpanzees of both sexes and almost any age other than infants or very young juveniles are frequently seen roaming calmly alone in the forest (Nishida 1968; Reynolds and Reynolds 1965). Even handicapped chimpanzees seem to be safe. Sugiyama (1968) saw a one-handed animal and also one which moved very slowly because of a crippled leg. This ape 'looked carefree,' and it was still with the group 1½ years later. Similarly, Albrecht and Dunnett (1971) report a group containing a mature female with a completely paralyzed arm, which had been seen by previous Netherlands researchers in this group 2 years earlier.

Albrecht and Dunnett (1971:117) suggest that, as is general among many other animal species in their study area, leopards and chimpanzees do not interfere with each other, but live 'in peaceful coexistence.' This may be in part because animals that leopards prefer as prey are numerous in the forests. Reynolds (1965*a*) mentions that Budongo Forest abounds with several species of small antelope, including the duiker, a favorite prey of the leopards and pythons. In more sparsely wooded areas, where food may be more difficult to find, the apes sometimes appear alarmed by the calls of lions and leopards (Itani 1979; Baldwin 1979).

The territoriality of wild chimpanzees

Anthropologists tend to use the term territory and home range interchangeably when referring to the area occupied by an individual or group, whether it is defended or not; whereas primatologists usually consider these to be different concepts. A range is defined as above, but when any part (or all) of this range is consistently defended against incursions by other conspecifics, the term territory is preferred. In this study, the term territoriality is used to describe the patterns of usage of either range or territory. Perhaps nothing so vividly demonstrates the usual social atmosphere of wild chimpanzees (and of the stressed Gombe and Mahale) groups than the way each utilizes their home range or territory.

All researchers agree that there are loosely structured local groups of chimpanzees. Reynolds and Reynolds (1965) refer to these as 'home range groups,' Sugiyama (1972) uses the term 'regional populations' and Nishida (1979) 'unit groups.' The Reynoldses' term home range groups is used in this study.

Wild chimpanzees, wherever they have been studied, are reported to live in fissioning and fusing home range groups of perhaps 60 to 80 animals (40 to 50 adults). Through familiarity with the forest and habit they tend to range the same general area of the forest. The home ranges of neighboring groups tend to overlap to a considerable degree, usually in a good feeding area which both utilize (see Figure 5). In fact, the Reynoldses emphasize that the 'home ranges' cannot be considered to be 'ranges' in the normal sense. Instead, there are 'areas of

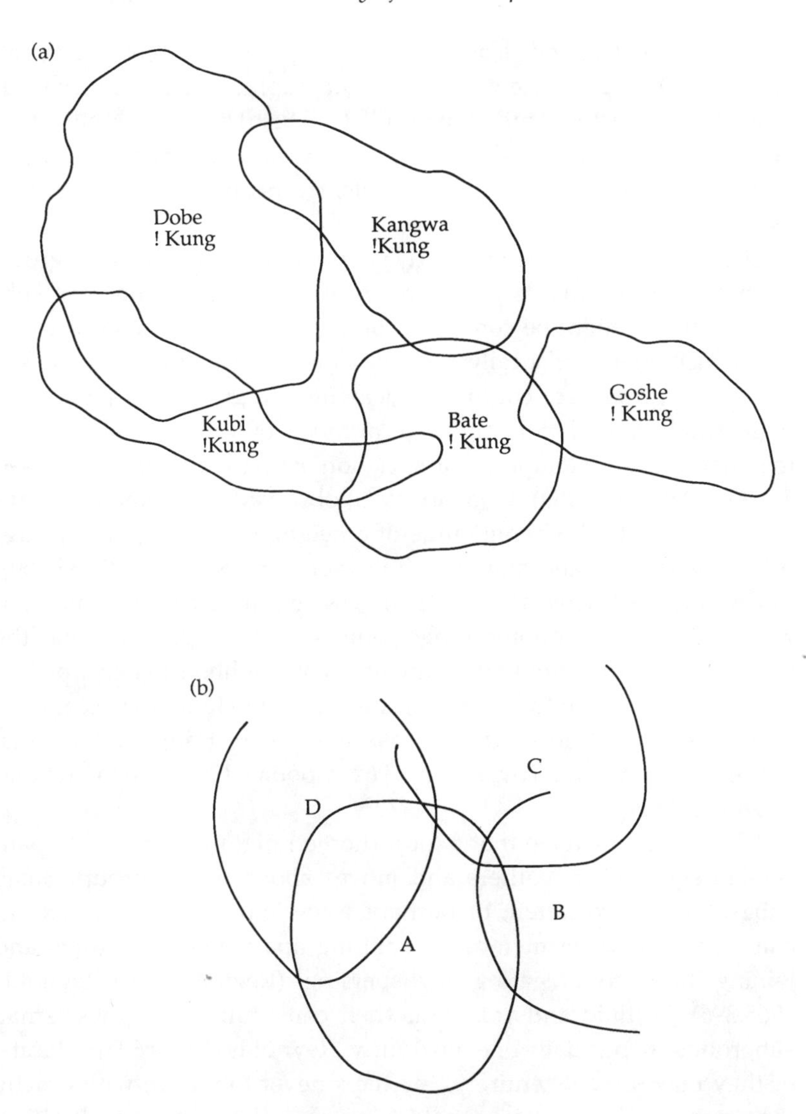

Figure 5. Overlapping home ranges of (a) humans (land use among the !Kung (adapted from Yellen 1976:55)) and (b) chimpanzees (adapted from Sugiyama 1973:77).

concentration,' familiar to certain groups of apes which have 'a higher frequency of interactions among one another than with bands beyond' (Reynolds and Reynolds 1965:400).

In the early (pre-1965) years, Goodall did not think that chimpanzees could be divided into separate communities. Because apes seemed to enter and leave the local group so freely, Goodall (1965*a*:456) thought it likely that only a geographic barrier would constitute a limiting factor. Among (immediate-return) human foragers, too, home ranges are open and overlapping, and are 'determined to a large extent by natural barriers, such as hills, rivers and ravines' (Turnbull 1968*a*:24).

Goodall's impression in the early years was that a home range group of chimpanzees consists of a number of small, very temporary subgroups which might be composed of any combination of age or sex, and which characteristically remain stable for a few hours or a few days. 'Individuals are continually leaving one group and joining up with another, and sometimes a number of groups move about together forming a temporary association of up to thirty individuals or so' (Goodall 1967*a*:289). Goodall (1973) observed that some chimpanzees moved into the home range of a neighboring group and peacefully traveled, fed and mated with its members. Reynolds (1963:100), too, reports that while some older or less venturesome wild chimpanzees seldom left the home range, others freely 'explore beyond the habitual areas and join up with members of neighbouring groups.' To repeat, there was nothing to indicate that wild chimpanzees are not free to come and go as they please without restraint by territorial borders (Azuma and Toyoshima 1962; Goodall 1965*a*; Reynolds and Reynolds 1965).

All researchers agree that a local (home range) group of chimpanzees rarely, if ever, gathers and moves about as one group. Small subgroups of two to ten, or perhaps a few more, form and reform, constantly changing members, 'splitting apart, meeting others and joining them, congregating or dispersing' (Reynolds and Reynolds 1965:396). So fluid and volatile is their constitution that these small subgroups change daily or even hourly, Reynolds (1965*b*:695) submits; or they may dissolve entirely, 'perhaps never to reform with exactly the same number or personnel.' Sugiyama (1972), Kortlandt (1962) and Goodall (1965*a*) make strongly similar observations. 'Individuals were free to join or leave a group at will, and the groups themselves often merged or split up' (Kortlandt 1962:132). Indeed, in the early years of her study, when she still followed the apes through the forest, Goodall (1967*b*:168) seldom saw the same two adults together 'more than once or twice in any given month.'

There is very close agreement as to the composition of these subgroups, which are viewed as being formed on a basis of free choice by each animal. Rank, age or sex do not come into the choice of traveling companions (Azuma and Toyoshima 1962; Goodall 1965*a*:430). It is mutual attraction, i.e. friendship, which seems to determine the composition of these groups. Sugiyama (1972:150) suggests they are made up of chimpanzees who share 'a feeling of well being and assurance.' Participation in any kind of aggregation also does not seem to be fixed or controlled in any way (Goodall 1965*a*; Reynolds and Reynolds 1965). There is a high level of individual autonomy. 'Chimpanzee society ensures the free and independent movement of each individual based on highly developed individuality without the restriction of either territoriality or hierarchy' (Sugiyama 1972:159–60).

Western sociologists tend to think of autonomy as being the ability of an individual to select personal values and 'to withstand social pressures for conformity,' and 'to act independently of the norms of an immediate group situation in favour of his (or her) own personal norms or convictions.'[6] A high level of personal and group autonomy is characteristic of foraging people (Marshall 1976; Woodburn 1982), but this sociological definition is not applicable. Among both humans who live by the immediate-return system of foraging and wild chimpanzees *autonomy* is a state or condition in which *self-organization and self-direction is expected as a property of the group*, even of a scattered group. It is a matter of the individuals interacting in the necessary roles and relationships being able to control and regulate their own lives. It implies the freedom to choose whether or not to be organized by others, by which others, whether or not to take part in any activity or relationship, to join or not to join, to come or to go. This structural element of autonomous adulthood is balanced by a strong need for the company of others. As explained in Part 5, this leads to a high level of 'independent conformity' which is usually in harmony with the norms of the group. A high sociality and the propensity to change companions combine to keep the element of self-direction from taking on an aura of 'me first'.

Characteristically, the small mobile subgroups of wild chimpanzees roam widely throughout their home range and the neighboring tracts (Goodall 1965*a*:455), usually hooting noisily and drumming on tree buttresses as they travel. The booming noise of their drumming could be heard by the Reynoldses at a distance of 2 miles (3 kilometers) through the dense forest. When the mobile apes find a good food source, such as a very large fruiting tree or grove, the vocalizing and drumming increase markedly, attracting other roaming groups and solitary wanderers, as well as alerting the more sedentary, the aged and the timid of both sexes and the mothering females. It was Reynolds' (1965*a*) impression that the calling leads to a general wavelike movement of any interested apes in the same direction, to swell the congregating larger group.

These large feeding groups of both sexes and of all ages tend to be very temporary, confined to 2 or 3 hour aggregations of smaller parties and individual animals that have come to a place where food is concentrated (Sugiyama 1972:146). When such a larger social group forms, the animals mingle with much excitement but without hostility. At first, the gathered apes display wildly. There is a great uproar of barking, hooting, wild screaming, swinging and shaking trees, stamping, dashing about, drumming on tree buttresses and so on, which appears to relieve initial tensions, as after a time the animals quieten, and begin calmer forms of social behavior: grooming, copulating, feeding and so on. Goodall (1963*b*) testifies that the period of uproar is very positive in tone. In her opinion, this behavior expresses 'merely excitement and pleasure', for 'with his highly emotional extrovert temperament, the chimpanzee likes to express his feelings in action' (Goodall 1963*b*:289).

The local people call these larger group gatherings '*ngoma*', meaning drum-dance, but, because of the high excitement and social interaction, Reynolds (1965*a*) felt that Garner's (1896) earlier-used local term kanjo, implying 'carnival', most closely described the social atmosphere or 'climate' of these gatherings. There is an interesting parallel to the behavior at Mbuti foragers' gatherings. Turnbull reports that they dissipate potential aggression in ritualized and contained acts of aggression and, in Leacock and Lee's (1982:143) words 'riotously humorous (rough) play.'

The 'naturalistic' observers generally agreed that there are friendly nonantagonistic social relations with the chimpanzees of the neighboring groups, even when apes from different neighboring groups congregate at common food sources (Goodall 1965*a*; Reynolds and Reynolds 1965:401; Itani and Suzuki 1967; Sugiyama 1972, 1973, Ghiglieri 1984). It was Reynolds' (1965*a*:159) impression that the carnivals occur 'when groups from two different regions met and joined, either to feed briefly together or to stay on a common feeding ground for a few days.' According to foraging theory, this would be optimal strategy for chimpanzees (see Part 4).

Typically, when these carnival-like gatherings taper off, some of the animals change groups, choosing to move about with a group other than that with which they arrived, for a time (Goodall 1965*a*; Sugiyama 1972; Nishida and Kawanaka 1972; Reynolds and Reynolds 1965). As among human foragers, there seems to be nothing to stop an individual from moving from one group to another, and from that to a third, as he or she wishes.

There is a general agreement among the unobtrusive field observers that the wild chimpanzees' social unit (local or sub-group) might be described as loose, fluid, flexible, ever changing in size and composition, being formed in 'pick-up' style, based on individuals' personal liking for each other, free choice and whim (Kortlandt 1962:132; Goodall 1965*a*:453; Reynolds and Reynolds 1965:396; Sugiyama 1972:160).

Because chimpanzee home ranges usually overlap quite extensively, Sugiyama (1972) suggests that there may be more design to the meeting and mingling of neighboring groups – more of a social bond between the home range units – than is usually recognized. It appears that, like human foraging groups, the wild chimpanzees extend fellow-feeling or, in anthropological terms, *social kinship*, to all of their species with whom they come into contact occasionally, yet fairly regularly.

There is no way of knowing whether chimpanzees are conscious of kinship with others, apart from the mother–child and probably the sibling relationship. But they need not be conscious of the device of social kinship, in order for it to function. The intense socialization that is typical of larger society carnival reunions suggests that these apes obviously do experience a 'connectedness' on a broad basis of generalized attraction and attachment to each other. As I show later, this generalized attachment is of value to them as good foraging strategy.

The behavior of the wild chimpanzees, and other aspects of their social structure, are examined more fully later in this book. However, enough has been advanced to indicate that the wild chimpanzees studied in earlier years lived in their natural habitat in conditions of relative safety and abundance. Food was sufficient; predation pressure was, in most locations, nil; human pressure – at the time – inconsequential. Social pressures were dispersed through a fluid pattern of fission and fusion and, as shall be shown, other smoothly operating social mechanisms; making aggression among adults unlikely and unacceptable, just as among the human foraging peoples – and for the same organizational reasons. Although, as we know, chimpanzees nowhere live in the undisturbed, uncrowded world to which they are adapted, there is unanimity of opinion that wild chimpanzees, everywhere that they have been studied, are not merely nonaggressive, nonhierarchical and nonterritorial, but highly positive in all aspects of their social life. Where they live relatively undisturbed lives, the adapted social mechanisms work. It is not that the wild chimpanzees are without problems – ecological crises, or social deviants whose behavior threatens the group – but that their society is in equilibrium; that is (by anthropological definition), the social system remains in a steady, stable state or returns to that state, if displaced.

Viewing the social groups of wild chimpanzees as being in a state of equilibrium does not imply a static state but a dynamic constant readjustment to balance as statuses, roles, age-graded ranks and circumstances change. It is a tenet of evolutionary theory that the various species of animals are well adapted to their particular natural environment; thus, in a stable habitat over millennia there is little change (Jolly 1972). The anthropological position is in general accord: if the society is in dynamic equilibrium, if the social controls function smoothly, if the members of a society feel strongly and harmoniously about their group's way of life, if the adults instill support for, and belief in, the society's organization in the young at an early age, then the established social organization will change very little, if at all, over time.

Equilibrium theory, based on the study of the small-scale simple human societies, has tended to be thought of as static. Meggitt (cited by Foster *et al.* 1979:326) points out that stability is taken to be the norm and change the problem to be explained. It is true that humans everywhere are encountering such rapid social change that rebalance models, such as the equilibrium concept, are no longer useful. However, in undisturbed nonhuman primate societies, stability of social system through constant readjustment is the *norm*, and *change* is the problem to be explained.

Because of the functional necessity of positive relations, the wild chimpanzees rarely have reason to demonstrate the negative emotions and behaviors of which they are capable, so evident in the behavior of captive primates and in the recent behavior of the Gombe and Mahale chimpanzees. In the smoothly functioning wild societies, because there is little to arouse aggression, such potentially disruptive behavior is not learned by the young through observation and imitation.

It is the argument of this study that the peaceful nature of undisturbed chimpanzee groups is not that they lack the biological (hormonal) basis for aggression – obviously they have such potential. But equally it is an error to think of a state of peace – as we have done – as merely a passive or null state of lack of aggression.

The social climate of the provisioned groups

Anthropologists who have carried out long-term studies of human groups find that the primary advantage of long-term studies is the opportunity for studying social change (Foster *et al.* 1979). No such use has been made of the Gombe and Mahale data pools, although evidence of extensive negative social change is found throughout the publications.

The evidence found in the Gombe publications is particularly clear. As suggested, for that reason the argument regarding negative social change as a result of feeding methods is based largely on data from the Gombe Research Center, although some Mahale data are used to fill gaps or to verify that similar social change is taking place among the Mahale apes also. When the many Gombe publications are read in chronological order, a radical detrimental change in social behavior so complete as to seem to be (as frustration theorist Maier (1961) suggests) a different set of behaviors, involving a different mind set, is clearly documented. By taking cognizance of the dates on which various behavioral data are reported in the publications, the reader can follow the commencement and contagion of negative social behavior as it spread from one aspect of chimpanzee life to another, step by recorded step, until the whole social structure of the provisioned Gombe group is distorted and malfunctioning. It is suggested below that direct competition by the frustrated apes for bait foods aroused a latent sense of possessiveness which is usually quiescent in indirectly competing wild chimpanzees.

Well-established findings from three disciplines are pertinent here. First, psychologists have established that tension is the required motivational state for competitive interactions. Second, animal behaviorists find that factors such as crowding and limits on food supplies, *which put animals not adapted to it into direct competition*, are major causes in disrupting a social system. Third, sociologists generally agree that the first aspect of a human society to be affected by an unusual level of stress and tension which the established social controls and mechanisms cannot disperse, is attitudes. Changes in attitudes usually lead to changes in behavior (and vice versa). Individuals affect changes of the system, and changes in the system cause further changes in the individuals who are involved with the system. The social atmosphere of the affected group or society may be rapidly and radically changed.

It was mentioned in Part 2 that, while complex human societies handle the mechanics of operation by dividing the system into separate spheres (political, economic, judicial and so on), there are no such divisions in the very direct (immediate-return) foraging societies. All aspects of the society function together as an interrelated and complementary whole. Disruption of one element or aspect tends to disrupt the entire pattern and to encourage disintegration in other elements of the system, or even of the system as a whole.

That tremendous tensions and frustrations built up in the Gombe apes (and local baboons) as they waited for the food boxes to be opened is testified to by both Goodall and Teleki, who, as a doctoral student and one of the earliest independent researchers, spent his first 12 months of field studies at Gombe in 1968–9. In his 1973 monograph, *The Predatory Behavior of Wild Chimpanzees*, Teleki describes the killing of an infant baboon by a particularly nervous and aggressive chimpanzee male, Mike, who became despot for a time in the stressed Gombe group.

The banana boxes had been filled the evening before, and the first box was sprung shortly after 7:00 a.m. when the earliest chimpanzees arrived. Only a few boxes had been opened when baboons of all ages began to stream into the camp. By 8:00 a.m. there were about 18 chimpanzees and 60 baboons, 'so the remaining boxes were kept shut in hope that at least part of the baboon troop would leave shortly' (Teleki 1973:62–3). The animals could smell the bananas, so were expectant. As the minutes passed and the boxes remained closed, several chimpanzees, particularly the adult males, became 'increasingly agitated (or frustrated). In spite of the growing tension, feeding is delayed a bit longer in order to avoid the chaos that would result from having more than 60 individuals of both species present,' Teleki (1973:68) reports.

During the morning referred to above, there were at least ten cases of interspecies aggression within minutes, and the intense excitement approached 'chaos' as more chimpanzees and baboons entered the camp from many directions. At approximately 8:20 a.m. aggression suddenly erupted between the chimpanzees, as one adult male attacked another nearby. The boxes were then closed by the observers and opened again 'when most individuals had calmed down considerably' (Teleki 1973:63). Sporadic feeding then continued throughout the morning hours as researchers attempted (by opening and closing the food boxes) to keep competition at a low level without creating undue frustration among the chimpanzees, Teleki explains.

'Possibly taking advantage of the ensuing uproar and general distraction' that went on that morning, Mike snatched an infant baboon from its mother's arms and fled bipedally, killing the infant through flailing it against the ground as he ran. His action elicited a burst of excited screaming from the apes and barking from the baboons, all of which converged on Mike's trail. Baboons and chimpanzees lunged and flailed out at one another; one male baboon leaped on Mike's back, snapping at his shoulder. 'By 8:22 all is relatively calm again, though there appears to be residue of tension,' Teleki (1973:63) comments.

By 1968, the Gombe chimpanzees (and also the local baboons) had been suffering similar 'tremendous tensions and frustrations,' sometimes several times a day, for 3 years. A high degree of correlation between frustration and an instigation to violence (in humans) is generally assumed by psychiatrists; hence, they interpret postfrustration behavior in terms of direct and displaced aggression or some substitute activity, such as regressive behavior. Teleki (1975:169) reports that there was 'some overlap' in aggression and predatory behavior patterns' in chimpanzees, 'for the facial expressions and movements made while killing prey are sometimes similar to those made in aggressive interactions.' He cites the compressed lips and the 'uplifting, slamming and dragging' movements during what he views as predatory action, but which Goodall (1968*a*) categorizes in the context of (frustration-induced) aggressive incidents. Teleki (1975:173) indicates that these predatory attacks occurred more often after the apes had consumed large quantities of bananas rather than before they had been fed. Of the 30 predatory episodes observed during 1968–9, 19 occurred when the chimpanzees were satiated with the fruit. Although Teleki (1975:148) indicates clearly that, in his opinion, hunger is not the stimulant for the killing of young baboons and monkeys, some form, of 'residual hunger' based on nutritional imbalance of a diet based too heavily on the supplied bananas 'may account for this apparent anomaly'. This may be so, but again it may be some residual tension or frustration that is the catalyst. It seems reasonable to assume that displaced aggression may have prompted Mike's snatching and killing of the infant baboon. There seems to be some indication that the psychological explanation may be at least part of the cause, in that the baboons do not respond to the apes as potential predators. According to Ransom (1981:291) the baboons did not flee from the chimpanzees and seemed 'totally oblivious' to changes in posture and movement indicating impending aggression among the frustrated chimpanzee, 'that were so obvious to human observers'.

The impact upon the apes of the human-controlled, totally unnatural feeding situations was grave in terms of the effect upon behavior, societal tone and relationships. The effects are clearly recorded by Goodall and other Gombe observers. Under the tension of waiting with others for the fruit – present, smelled, but inexplicably unobtainable – the Gombe chimpanzees became fiercely, directly competitive; and the radical change which proved devastating to the whole adapted social order began. As Maier (1961) suggests this happened as a response to prolonged frustration, a different – very negative – set of behaviors replaced the set of positive behaviors that structures the societies of undisturbed, wild chimpanzees.

The recent territoriality of the Gombe and Mahale apes

The territorial behavior of the provisioned Gombe and Mahale chimpanzees illuminates the new social climate, although it is just one aspect of an extensive, eventually all-encompassing, negative social change. To begin, it may be recalled that Goodall's (1965*a*) observations between June 1960 and December 1962 convinced her that the local, then-wild chimpanzee groups could not be considered as being separate communities because they freely and frequently united and mingled without any signs of aggression. Similarly, there was no territoriality in the sense of exclusive use of space among the Mahale apes in the early years (Azuma and Toyoshima 1962).

Right up 'to at least 1964' the local groups at Gombe associated freely (Goodall *et al.* 1979:21), but after that date, gradual division took place in the groups which came regularly to the Gombe feeding station. In 1968 it was realized that some of the animals which had always ranged further south than others were beginning to frequent the feeding station less regularly. The observers gradually began to view the habituated population as separate social communities, and to refer to the two groups as the 'Kahama' and the 'Kasakela' (feeding area) communities (Teleki *et al.* 1976:581). For clarity in the following argument they will be referred to as the 'split-off' (Kahama) and the 'camp' (Kasakela) groups.

In 1970–71 there was still some interaction between the two groups. By 1971, both were staying mainly in separate core areas, but every few days or weeks a party from one of the 'subcommunities' made an expedition into the other's area and reunited with their former comrades fairly peacefully (Bygott 1979:411). Yet, as Bygott reports it, their initial approach seemed aggressive in tone. Usually the visitors stalked the other group silently, then burst, displaying, into their midst, scattering the surprised group. After this bluffing display, however, the males usually began to groom and then to feed peacefully together. By 1972 the split-off group came no more to the feeding area, but remained permanently in the Kahama Valley near the southern edge of the park. Gradually the expeditions of groups of camp (Kasakela) males to the split-off group's location took on the character of border patrolling, and of avoidance behaviors (Goodall *et al.* 1979). Similar patrolling has been reported more recently at Mahale (Itani 1982).

Later reports indicate steady and rapid deterioration of social relations between the Gombe camp apes and their split-off neighbors. By the 1970s, the two groups are reported to 'occasionally indulge in displays of power as individuals hurl rocks and wave branches and even briefly attack one another' (Goodall 1973:11).

Animal behaviorists maintain that the striking feature about animal conflict 'is its conventional, non-injurious nature' (Maynard Smith and Price 1973:28). Aggression among animals under normal conditions is not employed with the objective of killing competitors, but of driving them away, in order to maintain normal spacing. Competing animals do sometimes seriously injure each other, but they tend not to use the 'weapons' they possess to their full potential, Maynard Smith and Price suggest.

In 1974, parties of camp males 'chased, caught and severely attacked' two of these former comrades (Goodall *et al.* 1979:34). Both victims had numerous serious wounds, and subsequently both disappeared. More attacks described as 'vicious' and 'brutal' assaults followed (see Goodall *et al.* 1979:35–41). In 1974, a family consisting of a mother and two daughters, one a young adult, the other adolescent, also former feeding station companions, were sought out, attacked and severely beaten. The adult daughter, who was in estrus, was forced, through threatening displays by the camp males, to accompany them north to their core area (Goodall *et al.* 1979:41). Eventually, after several such unprovoked attacks, the old mother was killed.

Groups of the camp males moved silently and tensely to the peripheral and overlapping areas of their range seeking out and seriously, or even lethally, attacking lone or outnumbered apes – not only their former companions who made up the split-off groups, but almost any chimpanzee not of their group who entered or came close to their home range. Goodall (1983*b*:34) reports that '15 "unhabituated mothers" have been seen to be brutally set upon by groups of Kasakela males.' Rather than seeming to be an attempt to drive outsiders away, it is the Gombe observers' impression that the aggressors were deliberately trying to kill their victims (Goodall *et al.* 1979:52). In Goodall's and her co-workers' opinions, although chimpanzee density within Gombe National Park probably has increased in recent years owing to incursions of human settlement pressing into the forest edges, there was not enough overcrowding to account for the violence of the attacks by the camp group apes. Ample wild food was still available. Yet, when the Gombe males encountered chimpanzees from neighboring groups, even those with whom they had long been familiar, they attacked and killed them – or, if outnumbered, retreated in haste. The overlapping areas of their home ranges became primitive war zones, and the interaction strongly reminiscent of primitive human warfare rather than the reunion carnivals they once were (see Vayda 1976). By the end of 1977 the Gombe apes had apparently sought out, attacked and, apparently, killed all of the group of ten males and females which had become a separate community.

Even prior to the Gombe groups dividing in 1972, serious attacks were seen to be made on 'stranger' chimpanzees. In all ten cases in which lone females encountered and attacked by the patrolling males could be seen clearly, the observers report that the female was severely wounded. Some of their infants were killed and cannibalized (Goodall *et al.* 1979:53).

Very recent Mahale reports are strongly similar. But before the similarities are cited, let us go back to the beginning of the studies at the Mahale Research Center. In the same way as at Gombe, the feeding station at the Mahale center was set up in an area where the ranges of two local groups of chimpanzees (the K and M groups) overlap. The observers found that for the first few years the two groups shared the overlap area, but they did so through a process of avoidance (Nishida 1979). The approaching group signaled their approach through vocalizing, and the other responded. Thereupon, one or the other group usually retreated from the overlap area without further contact.

In the 1970s, intensely agonistic relations developed between the males of the K and M groups. Between 1969 and 1975, all of the adult males of the smaller K group disappeared, one by one (Nishida *et al.* 1985). Most of the remaining K group females joined the M group. The Mahale observers strongly suspect on a basis of circumstantial evidence – the 'severe fighting' between the M and K group males and the Gombe precedent – that at least some of the vanished males had been killed by members of the larger M group (Nishida *et al.* 1985:288).

It is not necessary, in this study, to establish the reasons for these divisions. Gradually, however, many observers began to assume that the increasingly hostile relations between the males of neighboring communities are 'normally aggressive' (Wrangham 1979:481). Because of this widespread assumption, no importance as a cause for the separation seems to have been given to the high rate of very aggressive direct competition that took place over the bait foods. No cognizance seems to have been taken of the fact that the adapted form of feeding competition of chimpanzees is *indirect*, which, as already established, involves the separate, simultaneous seeking of the same essential resources (Mayr 1970). Direct competition is an important element in asocial behavior, just as cooperation is in prosocial behavior (Mackal 1979).

By 1986, Goodall accepts that chimpanzees are normally exclusively and fiercely territorial, far more violent in their hostility toward neighbors than most traditionally territorial animal species. This very honest and candid researcher (Goodall 1986*b*:488) reminds the reader that 'unfortunately' her data on territoriality are based on only the one habituated chimpanzee group (which split, for a time, into two), and that as yet we know nothing about the wild groups to the north and south of the habituated Gombe group. Nevertheless, she asserts 'that chimpanzees should be considered territorial' (Goodall 1986*b*:528).[7] She continues:

> But theirs is a form of territoriality that has shifted away from the relatively peaceful, ritualized maintenance of territory typical for many nonhuman animals, towards a more aggressive [violent] type of behavior. In the chimpanzee, territoriality functions not only to repel intruders from the home range, but sometimes to injure or eliminate them; not only to defend the existing home range and its resources, but to enlarge it opportunistically at the expense of weaker neighbors; not only to protect the female resources of a community, but to actively and aggressively recruit new sexual partners from neighboring social groups.

The rise of despots

In an early publication, Goodall (1965*a*) mentions that she did not consider the dominance concept very useful in considering chimpanzee social behavior in the pre-1965 years, because of the temporary nature of chimpanzee groups, the loose form of social structure, and because aggressive and submissive interactions between individuals were, at that time, infrequent. In addition, because the group constantly split up into small subgroups whose members changed constantly Goodall (1968*b*:333) suggests that 'all mature chimpanzees, whatever their status in relation to the group as a whole, may from time to time be the highest-ranking individuals [hence the leader] of a temporary association.' (See Silberbauer (1981) on leadership, Part 2).Goodall later abandoned this early insight for, in her later book for the general reader, (Goodall 1971*a*), she reports 'dominance contests' in considerable detail and, in so doing, reveals a distinct change of tone or social 'climate' in the fed group.

In the early years at Gombe Stream, Goodall (1971*a*) found that in the forest proper it was possible to follow a group of semi-habituated chimpanzees for hours without observing a single aggressive incident. After a total of 24 months of study of the Gombe apes between June 1960 and December 1962, Goodall (1965*a*:466) had seen 'mature males . . . fighting only on one occasion'. The few quarrels which provoked aggressive interaction were momentary squabbles; the longest such encounter that Goodall saw lasted 1½ minutes.

In an article based on data collected over 45 months between 1960 and 1965, Goodall (1968*b*:337) reports that in the period of 1 year she recorded 284 instances of chimpanzees attacking other members of the group. Sixty-six per cent (about 165) of the attacks 'were almost certainly due to the abnormal situation at the artifical feeding area,' therefore 'their frequency cannot be regarded as typical' (Goodall 1968*b*:337).

There were four occasions when wounding by biting was observed. As Goodall (1965*a*) reports only one fight and no wounding prior to December 1962, clearly all these woundings took place between 1963 and 1965, after artificial feeding had begun.

Only 10% of the 284 attacks could be classified as 'violent' (Goodall 1968*b*:342):

> Even attacks that appeared punishing to me often result in no discernible injury apart from the occasional wrenching out of hair. Other attacks consisted merely of a brief pounding, hitting, or rolling (by hitting or kicking) of the

individual, after which the aggressor often touched or embraced the other immediately.

In general, at the artificial feeding area disputes including fights occur more often. However, this provides an opportunity to study 'the mechanics of aggressive behavior' and 'changes in dominance' (Goodall 1971*a*).

In addition, the Gombe males began to use display behavior aggressively. Typically, display behavior among wild chimpanzees – the noisy attention-gaining hooting, thumping, charging, branch-dragging, brachiating and tree-shaking behavior which the apes engage in at time of excitement – is not used aggressively, nor does it seem to be directed against other chimpanzees. Kortlandt (1962) and Sugiyama (1972) suggest that the vigorous display acts as a nondirected, harmless catharsis, relieving tensions engendered by meeting others. Display also functions as a form of social solicitation, a way of attracting others, leading to associative behaviour, as suggested by Chance and Jolly (1970). Among the frustrated and tense Gombe apes, however, this characteristic behavior began to be used aggressively as a means of becoming a dominant animal (Goodall 1971*a*:117). Rank in a newly formed aggression-based hierarchy was gained and kept by use of tactics ranging from bluffing threat directed against another member to serious fighting. Goodall (1971*a*) writes of the powerful and very aggressive male Goliath, as the chimpanzee that she first identified as being the alpha or dominant male, until he was dispossessed as a result of a series of violent displaying contests with Mike. Both Goliath and Mike were males with a 'strong desire' for dominance (Goodall 1971*a*:123). Until he challenged Goliath and won despot position in the Gombe group, Mike had 'ranked almost bottom in the adult dominance hierarchy.' At one time Mike appeared to be almost bald from having hair pulled out during aggressive incidents with his fellow apes.

It is significant that both of these earliest despots were nervous, exceedingly aggressive animals, often shunned by many of the other apes, in the pre-1965 years. Writing on aggression, Hinde (1972:20) suggests that when 'mob violence' becomes socially acceptable in response to frustrating factors, it is those (animals or humans) who are insecure and most in need of status who are most likely to be the first to behave aggressively, and so gain hierarchical status.

Goodall (1971*a*) tells of the 'low ranking,' usually excluded Mike suddenly seizing two empty paraffin cans, charging, hooting, toward a group of males, bowling the two clattering cans ahead of him. This made an 'appalling racket', and the males scattered out of his way (Goodall (1971*a*:117–18). Mike repeated this technique on subsequent occasions and finally, first one, then another of the males began to approach him 'submissively' and to groom him (Goodall (1971*a*:119).

Goliath also began displaying more vigorously and more frequently than he previously had done and 'he too became more aggressive' (Goodall (1971*a*:119). Goodall began to doubt Goliath's sanity because he began attacking juveniles, and charging back and forth in so violent a manner that he would finally sit, panting from exertion, 'a froth of saliva glistening at his half-open mouth and a glint in his eyes that to us looked not far from madness' (Goodall (1971*a*:120). Fearing for her own safety, Goodall had a heavy iron cage made, into which she could retire when Goliath's temper was at its worst. Eventually Goliath was subdued by Mike's clattering displays, and Mike became the top-ranking or dominant male, Goodall reports. Mike 'continued to display very frequently and vigorously, and the lower-ranking chimps had increasing reason to fear him since often he would attack a female or youngster viciously at the slightest provocation' (Goodall (1971*a*:120).

Hinde (1972:21) reminds us that 'aggression as behaviour directed toward violence to others is not to be confused with merely "self-assertive" behaviour. It is not beneficial to society.' It is the excessive, apparently unprovoked attacks on the females and young that raise the first suspicions that this aggression-based dominance hierarchy is not the normal form of organization of chimpanzees, and that the alpha Gombe animal is a despot, using his power oppressively, rather than serving the group as a protective leader.

For many years the most widely used definition of dominance has been in terms of function. To be dominant was assumed to assure regular priority of access to the necessities of life and reproduction, usually postulated as food, mates and shelter. This functional definition does not necessarily involve aggressive means. Nevertheless, the negative connotation of the term dominance as involving aggressive means of gaining this rank is firmly fixed in the lexicon. This one-sided concentration utterly ignores the positive, protective, peace-keeping role which is the key function of dominant primates organized in a normal hierarchical order (Hall and DeVore 1965:56).

Except for recording Goliath defending the ape David Greybeard in a squabble with a baboon over bananas, and an instance of Mike defending an infant whose life he had endangered through his own aggressive actions a day earlier, Goodall does not report either of these dominant males taking on a society-maintaining, protective role. To the contrary, there are a great many reports of their attacking vulnerable members of the group without observable cause. The fact that these overaggressive animals could act in this way suggests a breakdown of the adapted control mechanisms (i.e. the structure of attention and application of sanctions) which, as we shall see, control deviants in an undisturbed group. It seems possible that their aggressive behavior toward the vulnerable females and young was frustration-induced scapegoating, and that these domineering males were simply aggressive, *uncontrolled* despots. The social controls and the form of leadership in the wild groups are taken up in Part 5.

Sexual relations among wild chimpanzees

There is unanimous agreement among naturalistic field observers that sexual behavior among the groups studied by unobtrusive nonfeeding methods was, in all cases, remarkably relaxed, non-competitive and amiable.[8] Copulation between wild chimpanzees is described in terms such as 'very casual', 'easy going' and 'unexcited' (Reynolds and Reynolds 1965:419, Reynolds 1965a:165; Sugiyama 1972:153; Ghiglieri 1984). The 'naturalistic' researchers agree that there is a complete absence of any sign of exclusive rights of males to estrous females and no competition for access, no attempt to monopolize them sexually, no permanent sexual partners and no signs of possessiveness or aggression between males in this, or indeed any other, situation (Kortlandt 1962; Goodall 1965*a*; Nishida 1968; Sugiyama 1972; Ghiglieri 1984). Most copulations, even those involving youthful, (adolescent) males are not interfered with by older or more dominant males (Kortlandt 1962; Goodall 1963*b*; Reynolds and Reynolds 1965; Sugiyama 1969, 1972; Albrecht and Dunnett 1971; Ghiglieri 1984). Estrous females are not constrained by the males in any way, but move about freely 'in the same loose way' as other mature wild apes (Albrecht and Dunnett 1971:118).

Copulation among the wild apes seems to be infrequent, although one is left with the question of how much of this behavior is missed under the difficult observational conditions of naturalistic study. On the other hand, Goodall, and Reynolds and Reynolds mention that unhabituated wild chimpanzees retreat rapidly on catching sight of them. Though incomplete samples, the behavior that the unobtrusive observers report usually took place before the apes became aware of the researchers. It should be normal behavior, unaffected by the knowledge of the presence of humans.

Reynolds and Reynolds (1965:418) observed estrous females in mixed groups 36 times. In 14 of these instances, the females copulated with one male and in the rest (22 instances) no male solicited, nor did the estrous female present. Goodall (1971*a*) reports that, even in post-1965 years, in the forest and away from the artificial conditions of her camp, actual mating, as opposed to grooming and just being together, seems to be comparatively infrequent. She writes of following a potential mating pair through the forest. The pair traveled together foraging, built the usual separate nests at dusk, and did not mate until the following morning. Then they mated but once 'before wandering off peacefully together to feed' (Goodall 1971a:191–2). Goodall (1965*a*:451) also notes that there was multiple mating in the larger groups situations, but seldom within small roaming subgroups. On one occasion, when a larger group was gathering, all seven of the males in the group copulated with the only estrous female present.

In a publication based on observations in 1961–2, Goodall (1965*a*) reports that copulation was solicited – and freely accepted or refused – by either sex. Rather differently, Albrecht and Dunnett (1971:119) report that by 1968 Goodall found that the Gombe males 'took the initiative' in 83% of the copulations 'or attempted copulations.' These observers point out that only about one-quarter of the copulations in their (categorically wild) group in Guinea were initiated by males. [Unfortunately, no numbers are given].

Sugiyama (1969:212) observed 19 copulations among wild chimpanzees, 14 of which were clearly solicited by females and 4 by males, which is similar to the percentage Albrecht and Dunnett (1971) report. However, radical differences or changes in the behavior and tone connected with the sexual aspect of Gombe chimpanzee life are more significant than numbers of copulations or which sex solicited most often, no matter what percentage of the total copulations or sexual solicitations researchers observed.

These reports indicate that, among wild chimpanzees, copulation is usually but not invariably a matter of female choice and male response to a female's postural sexual invitation. More important to the thesis of free choice being developed in this book is the observation that under natural conditions presentation or solicitation by either sex does not mean that copulation follows automatically. The wild chimpanzees of both sexes are amiable, perhaps sexually obliging, but certainly not nondiscriminatory. The evidence is that neither the male nor female will automatically 'service' any or all of the opposite sex. There are personal preferences. Sometimes the soliciting animal is accepted, although the team of Reynolds and Reynolds (1965), Sugiyama (1969), Ghiglieri (1984) and Albrecht and Dunnett (1971) observed receptive females present to an adult male and be refused. Sugiyama (1969:211) reports males 'requesting copulation' with an estrous female by grasping her buttocks and taking the mounting position. Sometimes copulation took place, but at other times the female stood erect, thus refusing the male's solicitation. There was, and is, no aggressive reaction on the part of a rejected wild male (Sugiyama 1969; Ghiglieri 1984).

Goodall (1965*a*), Sugiyama (1973) and MacKinnon (1978) suggest that conditions of social excitement seem to affect the amount of copulatory behavior, and that by far the majority of the matings probably took place during the large noisy excited gatherings of the larger society; the carnivals, as Reynolds calls them. On such occasions an estrous female might mate with two or three males within a span of only minutes.

Sexual relations in the Gombe group, post-1965

Yerkes and Elder (1936) found that, as among humans, the sexual behaviour of chimpanzees is significantly influenced by social and other psychobiological conditions, and social and other psychobiological conditions of wild and provisioned Gombe groups are very different. Let us look at the evidence.

There are significant and far-reaching changes in the sexual behavior of the provisioned apes, as the autonomous sexual choice of the wild chimpanzees (of both sexes) is increasingly replaced by coercive (male) pressure on the estrous females, for sexual access.

In 1965–6, when females were in estrus, Gombe males 'of high status' (i.e. the most successfully aggressive individuals) solicited copulation 'with a brief threatening courtship display' (MacKinnon 1978:76). The message was quite clear, MacKinnon comments, that if the female did not approach and submit, 'she was liable to be attacked' (MacKinnon 1978:76). Freedom of choice had begun to slip toward becoming the (coerced) choice of the directly competing, aggressively possessive males.

When Tutin studied the mating patterns and reproductive strategies of the Gombe apes over a 15 year period, including her own 16 months of study, ending in 1975, she distinguished three distinct patterns of mating. These she refers to as (Tutin 1979:30):

(a) '*Opportunistic,* non-competitive mating, when a receptive female may be mated by all of the adult males in the group.' Although by far the most observed copulations (73% (of a total of 1137) at Gombe, 70% to 90% (of a total of 383) at Mahale) were of the opportunistic type, relatively few pregnancies resulted from this type of mating. (This suggest a social function, such as greeting and tension release, hence I prefer the term *social mating* for this nonreproductive form).

(b) '*Possessiveness,* when a male establishes a special short-term relationship with a receptive female and may prevent lower-ranking males from copulating with her' (Tutin 1979:30). Nishida and Hiraiwa-Hasegawa (1986:169) point out that at Mahale the possessive pattern is a matter of a male, usually the dominant, using threats or aggression 'to prevent other males from mating' – rather than his gaining exclusive mating. Also, the dominant male seldom, if ever, goes away in consortship, which does facilitate exclusive mating. The Gombe and Mahale observers are aware that alpha or dominant rank does not yield 'alpha' reproduction rates.

(c) '*Consortships,* when a single adult male escorts a female away from the group and maintains exclusive copulatory access to her, as both of them take positive steps to avoid other chimpanzees' (Tutin 1979:30). Most pregnancies resulted from consortships (9 from 14 consortships). Thus, this pattern is taken to be the optimal reproductive strategy for males generally, and for females 'as they gave males the highest probability of reproductive success, and allowed females to exercise choice of mates' (Tutin 1979:29), which they did not always have, at Gombe, by 1972–5. When female choice is involved, 'the selection criteria appear to be social and caretaking abilities of the males and not their dominance status' (Tutin 1975:448).

The argument in this volume is that the evolved chimpanzee pattern of mating is free choice, based on indirect competition and the freedom of *either* sex to request, accept or decline mating. The possessive and consort forms are largely post-1965 innovations at Gombe, the development of which can be traced.

In her book *In the Shadow of Man,* Goodall (1971*a*) affirms that (after 1965) estrous females are constantly followed by a retinue of at least six males, a behavior not observed among the wild chimpanzees. Goodall (1967*b*:134) describes vividly the courting behavior of the six Gombe males soliciting an estrous female, as 'one by one they swing through the trees with exaggerated and stylized bounds and leaps before the final state . . . when they swayed and swaggered before her.' The same display behaviors may be either negative or positive, according to the emotive tone in which they are enacted. Indirectly competitive positive display, not directed at any individual, i.e. males 'showing off' their physical prowess – which may act to impress on the others their potential as protectors of the vulnerable – may attract estrous females. But the display behavior employed by the directly competing Gombe males as sexual overture began to take on a different emotive tone from the 'display to impress' behavior of the wild chimpanzees. In a publication 8 years later, Goodall (1975:138) indicates that 'most courtship displays have many components of aggression,' suggesting that a negative 'tone' of threat or coercion has crept into the males' sexual approach.

At Gombe, male 'request' increasingly became 'demand.' The female's freedom to choose her sexual partners was increasingly curtailed. Goodall (1971*a*) writes of a confident young female, Gigi, who invariably had a large male following when she went into estrus. Gigi would not accept the persistent solicitations of an aggressive male, Humphrey. After she had accepted other males, 'there would be Humphrey, his hair on end,' glaring, stamping and moving towards her, while she continued moving away (Goodall (1971*a*;186). Nevertheless, Humphrey often 'managed to get his way through dogged persistence.' Such persistence contains a strong element of coercion.

In post-1965 Gombe, of 213 copulations 'or attempted copulations' observed in 1 year, males now 'took the initiative' 176 times, females only 37 times (Goodall 1968*b*:361). Some females showing sexual swelling were mated repeatedly by virtually all adult males in quick succession several times a day (Goodall (1971*a*:87).[9]

None of the naturalistic field observers observed any male insistence, frenzied courtship, or attack on a female during estrus, or at any other time. However, in an article based on 15 years of post-1965 data, Goodall (1983*b*:35) reports that 'females often sustained deep gashes on their bottoms when they were attacked by males during periods of sexual swelling'. This kind of violence cannot be adaptive in a society in which a female breeds a number of times throughout a lifetime.

Such aggressive male sexual behavior imposes high costs on the females.

When, as in wild groups, a chimpanzee of either sex may solicit copulation, and either sex may accept or reject the sexual invitation as he or she feels inclined without penalty or repercussion, direct competition between controlling males, with its attendant tension and possessiveness, is precluded. The freedom of individuals of both sexes to accept or reject sexual solicitations enhances the easy sociability between males and between the sexes. It also makes possible ostracism from mating of individuals of either sex whose noncooperative, deviant behavior threatens the smooth functioning of the social system (see The negative sanctions, below).

Two years after the earlier-mentioned female Gigi had constantly refused the sexual solicitations of the aggressive male Humphrey, she seemed to Goodall (1971*a*:186) 'almost to prefer Humphrey to any other male.' Two years later Humphrey was the reigning despot at Gombe. It seems likely that in a society based on power aggressively imposed, wherein all members compete for dominance positions in the hierarchical system, its more vulnerable members would be safest in not resisting – or even in courting – the attentions and attendant protection of a link with the despot animals. Indeed, Yerkes and Elder (1936:38) found that when a captive female was sexually solicited by a threatening male 'the cautious, apprehensive or terrified female may present to the male whatever her physiological state.' When the generalized protective role of the group breaks down, there is a need for powerful individual protection. Although we cannot know her motivation, such reasons may explain Gigi's change of attitude toward the male Humphrey.

Although overt personal monopolization of estrous females is an exceptional pattern in the Gombe group, attempts at such individual possession on the part of despot males began to be observed between 1966 and 1969 (McGinnis 1979). Such attempts were seldom successful, because while a possessive male was occupied in attacking a competitor for copulatory access to a female, other males quickly took advantage of the opportunity to copulate with the female. Sometimes possessive males 'were seen to get around this problem' by charging and even attacking the female if she responded to the courtship of other males (McGinnis 1979:437).

Prior to 1965 at Gombe and before 1975–6 at Mahale, a pattern of consortship, an estrous female and a male going off together, away from the group, to travel as an exclusive mating pair for days or weeks, was uncommon. Prior to 1965 Goodall (1968*a*:219) records only ten instances of estrous females going off with a single male in consortship, away from the group at the feeding area, into the forest for three or four days. Since 1965, more incidents of such a character have been observed, Goodall (1968*a*) reports. At Gombe, also between 1966 and mid-1969, a type of 'abductive' or coercive consortship not recorded earlier, was observed (McGinnis 1979:431, 438). Males were seen to display threateningly and, if necessary, attack females who were hesitant in following them. In December 1972 'continuous (exclusive) escort service by most dominant males' was observed for the first time by Mahale researchers (Nishida 1979:113). 'Only recently', apparently about 1975–6, was 'safari' or consortships also observed (Nishida 1979). At least one Mahale male has been seen to 'hit and slap disobedient females' until they follow (Nishida *et al.* 1985:295).

McGinnis (1979:436), who writes of the sexual relations of the Gombe apes in terms of the 'copulatory rights' of the adult males to estrous females, suggests that estrous females appear to learn very quickly to follow the males' lead 'without much hesitation', as a result of 'behavioral shaping' by the males. The use of the euphemism of 'behavioral shaping' – and notion of 'disobedient females' – acts to soften or screen the reality, which is male threat and actual attack on estrous females who are unwilling to enter into a sexual liaison with them. Screens are not helpful to our understanding. In both provisioned groups, the once autonomous female chimpanzee has lost *her* sexual rights, which are the freedom of choice to accept or reject sexual solicitation.

Tutin (1975) suggests that not all of the Gombe consortships are coerced; most are by the female's choice and willing cooperation, gained through positive social and care-taking abilities of the males, and not through their dominance. For instance, a highly nervous young female, Pooch, showed a 'stable preference' for old Flo's charismatic adult son Figan in post-1965 years (Goodall 1971*a*:191). McGinnis (1979:435) noted that when the timid female followed Figan away from the group on extended consort trips, she did so 'with little or no coercion.' For one 6 month period, every time this female came into estrus the pair went off into the forest, away from the group.

It is understandable that a timid female would prefer the calmer consort situation. It is a solution to the problem of being the center of so much tension and aggressive competition. Voluntary consortships can be seen as a means by which a harrassed female may still choose a mating partner, rather than being forcibly chosen – which may explain the apparent increase in consort relations at Gombe, since 1965, and at Mahale, since 1975. There is not this need in the wild societies.

A 'gentle crucible': the infant experience in wild chimpanzee groups

As Poirier (1972:8) emphasizes, an animal's socialization, i.e. the sum total of its past experience, 'may be expected to shape future social behavior.' In other words, the individual is a 'product' of a given socialization procedure, fitted to live in a particular social environment. It is through socialization, through learned social responses and relationships with other members of the group, that chimpanzees (like humans) learn how to interact and behave within the social milieu. Any frequently encountered form of behavior, whether adaptive or maladaptive and destructive to the social order, may be passed on through socialization. Aggressive – or nonaggressive – responses and patterns may be learned through a sequence of observation, imitation and practice.

In the 1950s, John Marshall filmed an extensive record of the 'old ways' – at that time still holding – of the daily life of one of the last remaining immediate-return hunting–gathering groups, the !Kung San (people) of the Kalahari Desert. In a personal communication to E. R. Sorenson, Marshall referred to the !Kung's traditional foraging way of life, in a lovely and significant phrase, as a 'gentle crucible' (Sorenson 1978:12).

All the evidence on development of young chimpanzees in wild social groups indicates that the infant experience is highly positive, in a socioemotive atmosphere of gentle affection and acceptance from all. Kortlandt (1962:130) comments that chimpanzee 'children' lead 'a wonderful life.' He reports that they are pampered and free by human standards; but despite their freedom, the young apes are well behaved and obedient (Kortlandt 1962:130). They, too, grow up in a gentle crucible.

By Goodall's (1965*a*) early classification, infants are the young from birth to about 2½–3 years. Once a young ape is independent of its mother for feeding, transportation and sleeping, it is considered to be a juvenile (from 2½–3 to 7–8 years). In the pre-1965 years, when

Gombe infants first began to venture a short distance away from their mothers and to play with other infants and older children (at from about 4 months to 1½ years), the play was always 'slow and gentle', Goodall (1965*a*:432, 1963*b*:287) reports. Kortlandt (1962:130) observes that juvenile chimpanzees were allowed 'to touch even the smallest infants and gently investigate them.' Juveniles 'are very gentle when playing with infants and are always quick to run and help them' if they get into difficulties (Goodall 1965*a*:460). Older infants often sought out adolescents for play; on many occasions Goodall saw older infants hurl themselves on to resting adolescents of both sexes, pummeling them. Sometimes the adolescent responded by sparring with the infant, or it 'merely patted it from time to time.' Goodall (1968*a*:259) saw adolescents become 'slightly boisterous' with infants on four occasions, but they 'and juveniles were normally gentle with younger animals . . .'.

Reynolds (1965*a*:178–9), too, reports a 'delightful panorama' of gentle child play which he observed, during one interaction of an infant and a juvenile ape of about 3 years. Later a different infant, a juvenile and a female adolescent 'all joined in playing a chasing game,' and when their mothers decided to move on, the adolescent female 'popped one of the infants beneath her and carried it in place of the mother.' Goodall (1968*b*, 1971*a*) reports that adult males usually reached out to pat, tickle or hug gently infants who approached. Similar playful interaction between Mahale infants and adult males is reported by Nishida (1979). There is no record of the kin relationship of the actors involved in these exchanges. However, according to Kortlandt (1962), there was affectionate acceptance of chimpanzee infants by all members of the social group.

In the early 1960s, Claire and W. M. S. Russell undertook a survey of the reports of all field studies of primates then available. This survey, and a remark by Kortlandt (personal communication, cited by Russell and Russell 1972), led them to the conclusion that there are two sex-based parental roles in nonhuman primate groups. There is a kind of protective *social* parental behavior on the part of the males, generalized toward all dependent young, and an intimate form of parenthood on the part of the females, i.e. a concentration on nurturing of her own biological child (Russell and Russell 1972:56–7). As Kortlandt (1962) suggested to the Russells, there are *multiple social fathers* (of the young of the group) and *individual biological mothers*. This is an enormously important insight, which will be referred to again in Part 5, as it helps us to understand the adapted social structure.

In a 1965 publication, Goodall drew on observational data obtained during her first 4 years of study of the then-wild chimpanzees at Gombe, to report on mother–infant relations. She noted (Goodall 1965*a*:458) that wild mothers are generally extremely solicitous of their infants. For example, when a young one is under 6 months old, the mother 'warns it when she is about to move off by pressing it gently to her body.' When it becomes older she signals her intent to move by reaching out to touch it or by simply gesturing towards it. 'The infant at once goes to her and holds on.' Kortlandt (1962) also reports this silent tactile infant–mother communication and emphasizes the offspring's quick, always obedient response to its mother's directions. Another facet of 'child' behavior which strongly impressed Kortlandt (1962), and also Albrecht and Dunnett (1971), was the silence of the wild chimpanzee young. 'They never whined or whimpered,' states Kortlandt (1962:130), 'and they always obeyed their mothers at the first hint, . . . infants in zoos behave rather differently in this respect.'

Goodall singles out males as playing particularly frequently with the infants more than did unrelated adult females. This would be so. Male paternal interest is generalized. The male chimpanzees, particularly the adolescents and the younger adult males, were very good-natured with infants, she reports (Goodall 1965*a*). Wild females are good-natured with the young also, as Kortlandt testifies, although their main concern is their own offspring. Infants could even 'occasionally invite themselves to share a meal with a strange mother and child,' he observes (Kortlandt 1962:130). Reynolds and Reynolds (1965), too, occasionally saw mothers caring for another's child. One particular mother had one infant one day, and the day following she was caretaking two. In general, as Albrecht and Dunnett (1971:45), observe 'most adults, both male and female, were very tolerant of young ones.' Social interactions between infants and other members of the group are frequent and, typically, highly positive in affective tone.

Once an infant is ready to venture a little apart from its mother, it is often carried by older infants, juveniles and adolescents (Goodall 1963*b*, 1965*a*; Reynolds and Reynolds 1965; Kortlandt 1962). Hence an infant frequently (and surely confidently) approaches individuals joining its group to greet them. Following Mason (1964), Goodall agrees (1968*b*:372) that the wild chimpanzee child is usually surrounded from infancy by individuals whose appearance and reactions to him (or her) are similar to those of the mother, i.e. caretaking, 'thus providing seemingly optimal conditions for the generalization or extension of filial attachments.' The young ape therefore learns by observation and practice, in a stable context of protective and affectional bonds from numerous caretaking adults of both sexes. As Goodall (1968*b*) suggests, generalization of filial attachment to the group as a whole is an easy next step.

These are the experiences of the human forager's child also. In a statement which unwittingly echoes Mason (1964) and Goodall (1968*b*), Turnbull (1978) comments that when a Mbuti infant begins to explore, to crawl or walk all around the camp, the baby learns that if some difficulty occurs she or he can expect to be comforted, but sometimes by a different 'mother.' Turnbull explains that the mothering individual may be an old man or young girl. Such experience enlarges the child's concept and experience of being mothered, and her or his sense of security, Turnbull (1978:176–7), continues, for now the child 'has a plurality of mothers and safe territories' and the first lesson in true sociality. The behavior Turnbull refers to as plural mothering is in essence that which Russell and Russell refer to among chimpanzees as social (generalized) fathering.

Although in human foraging groups the nuclear family is strongly bonded, all people, 'particularly men', play with children, teach them and discipline them 'almost without regard for the degree of (blood) relationship' (Turnbull 1978:292–3). It seems that, along with their special love for and parental role responsibilities toward their own children, the men – or, in general, the childless adults – in foraging societies also take the generalized social parent role that Kortlandt (1962) identifies among chimpanzees, while the mothers tend to retain the intimate biological parenting role.

When the human forager's child feels totally secure and 'at home' in this wider sphere, she or he is inclined to move on, to experiment with the next sphere of learning. Surely it is a similar atmosphere of generalized protective parenthood that normally permits most infant chimpanzees in the second year of life to wander about happily quite far away from their mothers (Goodall 1971*a*).

Upon weaning, at about 3 years of age, the Mbuti child may expect to be fed and protected by any family. By this age, they may 'equally expect to be disciplined by any adult' (Turnbull 1965:305). Similarly, de Waal, (1982:165) who observes that the adult chimpanzees in Armhem zoo are 'incredibly tolerant of infants' added that, as the infants grow older, more is expected of them. Adult acceptance in either species is increasingly based on the child's learning the social behaviors expected of his or her advancing age. The whole process of child training among human foragers generally is informal and emphasizes the child's responsibilities and connections to the group as a whole. This emphasis on socialization toward a generalized connectiveness to the group is both a goal and a value of human foraging peoples. It is probably a natural result of the parental roles, rather than a conscious goal, among chimpanzees.

Bowlby (1973:24) maintains that a person's childhood experiences greatly influence his or her expectation of finding someone who will supply a secure, supportive personal base, and the ability to enter and maintain a trusting, mutually rewarding relationship' with such an individual. In order to do so successfully, a person requires a healthy mixture of self-reliance and normal dependence, Bowlby suggests. Such people all report the same developmental background: 'a stable family base from which first the child, then the adolescent and finally the young adult moves out in a series of ever-lengthening excursions' when ready. Autonomy is encouraged, but not forced; thus each step follows the previous one in easy stages. Looking back to the evidence of the experiences of wild chimpanzees, and human foragers' children, it seems that the young of both species have the childhood experience that best facilitates the essential mixture of dependence and confidence which makes easy the frequent exchanges of role, and makes possible egalitarianism coexistent with a leader–follower relationship.

The Gombe infant experience

With the change in behavior, organization and affective tone of the Gombe society, the socialization experience of the young apes changed. Accepting aggression as a normal part of the network of social relations that structures chimpanzee society, Goodall (1986*b*:353) suggests that 'it is helpful to an understanding of the multi-faceted function of aggressive behavior to first look at the role it plays in infant socialization.'

Unlike the gentle, affectionate (generalized) caretaking response from each and every member of the wild society, the infant and juvenile chimpanzees at Gombe (post-1965) experience an uneven and often unpredictable amount of tolerance – ranging from affectionate acceptance, to simple impatience, to rough aggressive play, to life-threatening or murderous attack. The young, like the adult females and other vulnerable individuals, are often scapegoats for powerful males' frustration. While the secure wild young learn trust and a generalized affective affiliation with all members of the group, particularly the adult males, the post-1965 childhood experience of the Gombe young teaches mistrust, fear and avoidance of others other than their mother. That the Gombe young become insecure adults is suggested through the evidence of a prolongation of dependency on their mothers, and an accompanying delay in social maturity, as we shall see.

Unlike her earlier reports of gentle play between infants and older Gombe children, in a 1968 article Goodall (1968*a*:287) reports that play contacts of 2 year old infants with adolescent or even mature males 'were often rough and boisterous.' 'When frustrated in a feeding situation', older chimpanzee children direct a type of 'aggressive play' at others (Goodall 1968*a*:262). By the 1980s Goodall writes that 'the child assimilates a great deal about aggression ... during play sessions,' as play often ends in aggression (Goodall 1986*b*:355). In this way, the child 'develops skills in fighting' which will enable him to compete in a society in which fighting (for his 'rights', and dominance rank) 'may be important' (Goodall 1986*b*:355).

Generally, mature males were still tolerant of infants in their second year, but towards 3 years of age 'infants were sometimes pushed away rather more roughly than before during feeding and copulatory situations' (Goodall 1968*a*:287). Whether or not this change from unconditional tolerance to conditional tolerance is the quite usual one of expecting more, in social terms, of the maturing young, we cannot tell. However, pushing the young away from food is not characteristic among wild chimpanzees.

Unlike the usually tolerant wild mothers, Gombe females with infants 'probably directed more aggressive behavior towards [others'] 2 year old infants than any other age/sex class', Goodall (1968*a*) continues. Similarly, Nishida (1983*a*:29) reports that, of 12 recently observed interactions between mothering females and infants other than their own, 10 were abusive. Gombe mothers no longer permit inspection of their infant by these slightly older babies (as do the wild chimpanzees). They often threaten the older infants when they 'persistently' try to touch or play with their small infants, when they try to share the female's bananas, or when a squabble breaks out in a play group involving the female's own child. Sometimes the mother of a roughly treated child tries to distract the unruly playmate by tickling or otherwise playing with it; or, interestingly, at other times she retaliates by hitting the mother of the too-rough child (Goodall 1968*a*:236).

The fundamental role of the adult chimpanzees is parental protection of the young. The stressed Gombe chimpanzees of all ages and both sexes are still interested in, and often show affectionate care and indulgence toward, the very young. Although the social atmosphere began to change in the Gombe groups as the animals reacted to the frustration and competition over withheld bait foods, males still usually fill the male parental role of generalized protection. But there has arisen an ambivalence, a threat to the young from within the ranks of the normally protective 'fathering' males. Goodall (1968*a*:286) reports that a number of times she saw mature males in the protective role draw an infant to them, while mildly threatening humans who approached too close. However, she reports a great many paradoxical situations, such as seeing a 'high ranking' male run to pick up an infant of about 1 year of age who 'fell from his mother when she was attacked by another male.' Under stress, at Gombe, the male became both protector and threat to the more vulnerable (weaker, smaller, elderly, handicapped, timid) animals, including the infants.

The generalized form of social parenthood and its distortion at Gombe is exemplified by Goodall (1971*a*:153) when she writes of the behavior of a resting group of males toward a venturesome, very young infant whom she dubbed Goblin. On this occasion, the infant toddled over to the despot male Mike, who reached out and patted the infant's back very gently. Goblin then moved away, tripping over a root and falling. Immediately, a juvenile female 'hurried over and gathered him up, holding him close for a moment,' until the infant pulled away and again wandered off. He then tried to climb a small tree stump, but halfway up he lost his grip and would have fallen had not David Greybeard, who was watching, quickly reached out and steadied the infant. Goodall (1971*a*) comments that she was struck by the harmonious scene. [I am struck by the resemblance to the highly positive social parenthood experience of Mbuti infants, described earlier].

However, within a short time more chimpanzees could be heard approaching and Goodall, noting that Mike's hair was raised and that he was beginning to hoot, realized that he was about to display. So did the apes, she remarks, who prepared to join in or dash out of the way –

all, that is – except the infant Goblin. This 10 month infant (born in September 1964), who was barely able to walk, 'seemed totally unconcerned and, incredibly,' began to move towards Mike. Mike began his charge and, as he passed, seized the infant up 'as though he were a branch and dragged him along the ground.' The male 'flailed' Goblin against the ground, which is exactly the technique he used to kill a baboon infant that he snatched from its mother's lap during one of his 'frustration displays' at the feeding area, when he had not yet obtained bananas (Goodall 1968*b*:190, 289). Goblin's mother Melissa threw herself at Mike, and although she was badly beaten by the aggressively displaying male, she did succeed in rescuing the screaming and terrified infant as Mike dropped him to the ground (Goodall 1971*a*). While Mike attacked the mother, an old male (Huxley) had snatched up Goblin from the ground. Goodall feared that he too was going to display with the infant, 'but he remained quite still, holding the child and staring down at him almost, it seemed, in bewilderment' (Goodall 1971*a*:154). When Goblin's 'screaming and bleeding' mother approached, Huxley set the infant down whereupon he leapt to his mother.

Some psychologists hypothesize that frustration results in a level of tension which produces irrational behavior, the only aim of which is the reduction of tension. Consequently frustration reactions are often not adapted to the requirements of the situation. Expressing pent-up frustration through some action, even though not goal-oriented, does help to dissipate tensions; and 'as long as aggression is relieved through harmless channels, the condition is not aggravated' (Maier 1961:129). However, this kind of treatment of infants, which is a reversal of the normal protective parental role, can be assumed to disturb further rather than soothe the watching, as well as the participant, chimpanzees.

Some time later Mike demonstrated the normal form of protective parental role of the male (or childless) adult chimpanzee, in an incident which also involved the same young infant, Goblin. This time one of the males suddenly charged and briefly attacked one of two mothers who were sitting, mutually grooming, while their infants played nearby. When the male attacked, both females, one of whom was Goblin's mother, instantly raced toward their playing infants. Goblin's mother Melissa, who got there first, in error scooped up the wrong infant and ran. The other mother snatched up Goblin, realized the mistake, dropped him and ran after Melissa and her own screaming child. 'Goblin was left alone, his face almost split in two by his huge grin of fear,' writes Goodall (1971*a*;155). Then Mike ran up and this time 'his behavior was completely different.' The male very gently picked up the terrified infant and carried him for some distance. 'When the infant struggled to escape, Mike put him down', but stayed with him for ten minutes, 'threatening or chasing off other chimpanzees who approached too closely' (Goodall 1971*a*:155).

Actions such as Mike's, endangering the life of infants, are not isolated or unusual at Gombe in recent years. Other male chimpanzees were seen using infants as 'display objects' in the same context (Goodall 1968*a*:190), and a number of infants have been injured. Goodall notes that a chimpanzee that has been attacked by a more aggressive individual or that has been frustrated in its attempts to obtain bananas due to the presence of a 'social superior,' might threaten, chase or actually attack an individual subordinate to itself. 'When mature males attacked infants or juveniles it was usually in this context' (Goodall 1968*a*:274).

Goodall (1971*a*) tells of an orphan, Merlin (born 1961), who often advanced, bobbing submissively to mature males just prior to their charging (frustration) displays and, as a result, was frequently used as a 'display object' and thrown, or flailed against the ground. Time and time again this motherless infant 'was dragged or buffeted by displaying males because he ran toward them instead of away,' (Goodall 1971*a*:222). As a result there was 'a marked deterioration of Merlin's social responses.' By the time Merlin was 6 years of age his behavior had become abnormal, 'extremely submissive' to adults, and 'extra aggressive' to other infants (Goodall (1971*a*:222).

Goodall (1968*a*:287) also tells of the well-mothered Flint – a son born in 1964 to old Flo – at 2 years of age going in normal wild infant fashion to greet an approaching mature male, apparently oblivious to the fact that the male was rocking slightly and had his hair erect. Flo rushed to intercept, but was 'too late to prevent the child from being dragged over the ground and flung a few feet into the air during the male's charging display.' Later that day Flint ran and clung to his mother when the same male approached. Subsequently, however, Flint continued to disregard signs of frustration or aggressiveness in mature males, and his mother continued to have to snatch him from potentially dangerous situations. By the age of 3 years, Flint had apparently learned to recognize signs of aggression by males and 'on several occasions was seen to run and cling to his mother,' looking apprehensively in the direction of an aroused male. Other Gombe infants also often hurried back to their mothers when they heard adult members of the group scream or bark, Goodall observed. She reports (Goodall 1973:3–4) that, in the post-1965 years, if a young chimpanzee of 6 or 7 years became accidentally separated from its mother this resulted in 'obvious distress.' The child began to whimper and then to scream, scanning the countryside in all directions until it found its mother.

The behavior of wild chimpanzee children is very different. Reynolds (1965*a*) and Nishida (1968) report occasionally meeting confident 3 and 4 year olds moving alone in the forest. Albrecht and Dunnett (1971:44) assert that chimpanzees at this early age are 'already remarkably independent.'

The Gombe young were also learning to use aggression in their interactions with others. Indeed, aggressive and submissive patterns, 'become frequent during the second year of infancy,' according to Goodall (1968*a*:287). Goblin was but 15 months of age 'when he was first seen to stamp toward and hit a conspecific – a juvenile female,' writes Goodall, adding that the infant showed full hair erection, and that the context (a squabble) showed that the display was aggressively motivated. Almost a decade later Goodall (1979:603) reports that the very aggressive Goblin was showing 'signs of becoming the next alpha male.'

Psychologists Zimbardo and Ruch (1975:617) pose the rhetorical question: if you want an individual (human) to become very aggressive, what would be the ideal learning situation?

> First you would certainly want an adult model, and since children are dependent on adults, it would be good to have the learner be a child. You would want to make sure that the aggressive adult would be noticed by the child and that the child would be emotionally aroused. Both of these conditions are satisfied by having the adult punish the child – a child is certain to notice aggression, if he or she is the target of it . . . . The adult model, in turn, should be nurturant and should have rewarded imitation in the past. Thus parents would be an ideal choice.

Unlike among immediate-return foraging people and wild chimpanzees, aggressive models are readily available to stimulate Gombe young and adults to violent actions.

Not only the young, but the adult females also suffered, as a result of the dichotomy in the male apes' role behavior; and the paradoxical actions of adult males suggest that they, too, suffer confusion of feelings as a result of this behavioral ambivalence. Goodall (1971*a*:155) watched a hugh male, Rodolf, 'pound and drag an old female during one of his displays whilst her infant clung screaming beneath her; then, almost before he had stopped attacking her, he turned around to embrace, pat and kiss her.' When mothers of infants are attacked, they do not usually attempt to escape but remain in one place crouched over their infants, Goodall (1967*a*:311) attests. In this situation, too, the infant also is most certainly threatened and frightened, if not hurt.

Recent reports from Mahale Research Centre are that the males display similar 'ambivalent reactions,' apparently not toward females born in the group, but to immigrant females who had joined the group up to 3 years earlier (Nishida *et al*. 1985:295). These researchers report that the males bluff and chase these females at one moment, and groom and even protect them from other aggressive individuals at the next. As at Gombe, twice an incoming mother–infant pair from K group was severely attacked and would have been killed by the M group males, had not the observers driven off the attacking males. It was the observers' impression that these attacks were directed against the females, not their infants.

In attacking females and infant members of the group, the Gombe males are exhibiting a reversal of the usual generalized protective role. But, according to Russell and Russell (1972:57), the male primate's generalized social parental role (which may entail attack) is more easily reversed into hostile behavior under stress than is the intimate parental behavior of the biological mother. Goodall's (1986*a*) data indicate that the stressed Gombe mothers still try to protect their offspring against attack.

Between 1964 and 1967, 13 infants were born in the Gombe group, four of which died before their first birthday. The causes of the infant deaths are not known, but Goodall (1968*a*:252) notes that female chimpanzees are quick to leap out of the way during the 'charging displays' of the males or other violent social activities. She suggests that one cause of infant mortality may be injuries caused by falling from the mother. When a mother must run from the displaying males the situation is potentially very dangerous for a newborn offspring, Goodall points out, as the very young chimpanzee infant is not able to grip securely to its mother's hair. The mother must support it physically for several months after birth and so is severely handicapped in movement.

Mothers of even 'the smallest' infants, in the little-disturbed wild groups that Kortlandt (1962:130), observed, permitted other apes, even quite young juveniles, to touch and 'gently investigate' their infants. Very differently, in her 1971 book, Goodall makes a general statement that babies under five months of age 'are normally protected by their mothers from all contact with other chimpanzees except their own siblings' (Goodall 1971*a*:149). Gombe mothers are 'possessive' about their infants, Goodall (1975:135) explains. The Gombe mothers have ample reason for this possessive–protective behavior in a society where aggressive interaction between members sometimes involves infants as hapless victims.

Adaptation is for behaviors that lead to reproductive success – not the number of times an individual mates, but the number of their offspring, raised to adulthood, that can pass their genes to a next generation. When a primate society is exposed to any kind of stress, the progeny are particularly vulnerable. Under stress, parental treatment of children is affected, running all the way from 'neglect to competition, domination, attack and killing', (Russell and Russell 1968:268). When the young grow up, they tend to have their own parental behavior affected negatively and so the original stress can be passed on to a third or a fourth generation.

Infanticide and cannibalism at Gombe and Mahale

Infant mortality rates in entirely unhabituated groups cannot be calculated. However, from 1976 to 1983, Sugiyama (1984) studied the population dynamics of a group of chimpanzees at Bossou, Guinea, which by the definition used in this study are *wild*. Although they have become habituated to the presence of observers through repeated studies since 1976, the habituation has been achieved through familiarity, without provisioning or major interference with their way of life. The infant mortality rate among these Bossou chimpanzees is clearly lower than those at Gombe and Mahale. At Bossou the infant mortality rate (from birth to 3 years), based on eight infants born between 1974 and 1979, is 0.06. On the basis of 14 infants born between 1972 and 1982, the mortality rate at Bossou is 0.18. According to Sugiyama (1984:395), Gombe infant mortality rates for the same age group were 0.27 to 0.33. Mahale rates were 0.3 to 0.5 in the K group and 0.32 (1 year) in the M group.

From the anthropologist's perspective, the birth rate is usually seen as being adaptive to the normal rates of mortality. By this criterion, the long birth interval of 4 to 5 years and the pattern of single births among chimpanzees seem to indicate a normal low rate of mortality. The high percentage of infant mortality at Gombe and Mahale cannot be adaptive in the slow-breeding species.

During a 1966 epidemic of polio among the Gombe apes, 33% of the infants born, died. From 1971 to 1973, no Gombe infants died. However, in 1975–6 infant mortality arose to 83.3% (Goodall 1977). (See table II (births, miscarriages and infant mortality since 1965) of Goodall (1977:274)). Of eight infants born within the Kasakela (feeding area) group in the 3 year period from 1974 to 1976 (four to immigrant, four to resident females), all but one was known to be killed, or to have vanished, within the first month after birth. Four of these infants were killed and cannibalized by one adult female of the group. No other Gombe adult deliberately killed young born into the group, although groups of Gombe males killed, and both sexes fed on, the infant offspring of 'stranger' females.

In all, a total of nine infants, eight males and one female, is known to have been killed by Gombe and Mahale males (Nishida and Kawanaka 1985). With the exception of one case (which will be outlined shortly), observed by Suzuki (1971) in Budongo Forest, no naturalistic observers reported any evidence, or even suspicion of, infant-killing by wild chimpanzees. This, of course, is negative evidence. Still, as we know, Goodall (1965*a*,*b*) reports that (in the early years) juveniles and infants frequently ran toward incoming apes, including 'strangers', to greet them. This type of observed behavior does not suggest a habitual or necessary guarding of the wild young from the threat of attack by other chimpanzees.

Unlike the Gombe apes, the Mahale males have been seen to kill and cannibalize both infants of immigrant females and also infants born in their own group. Of the 22 infants known to have been born in the M group between May 1979 and 1982, 7 (31.8%) did not survive the first year (Hiraiwa-Hasegawa *et al*. 1984). The cause of the deaths of two of the Mahale infants is unknown; two died of disease; one infant male is known to have been killed and eaten by an adult male of its natal group. Circumstantial evidence suggests that two other male infants met the same fate. More recent reports from Mahale raise the totals. Now four cases of 'within-group cannibalism' in which adult males kill and eat the infants of females belonging to their own unit-group have been observed (Nishida and Kawanaka 1985:274). Further, five times mothers have appeared without their infants and were not carrying their bodies, as they usually do, when an infant dies. One of these females carried multiple severe wounds on her face and back, so cannibalism is suspected in these disappearances (Hiraiwa-Hasegawa *et al.* 1984; Nishida and Kawanaka 1985). In all four observed infant killings, the victims were male and 'very probably' offspring of the cannibals or their relatives (Nishida and Kawanaka 1985:274). The mothers, in every case, were ex-K group females who transferred to M group 1 to 5 years before their infants were killed.

Mammals specialized as predators do not normally prey on conspecifics. Cannibalism is rarely seen among wild mammals and was not, until 1967, observed among *wild* chimpanzees, when Suzuki (1971) saw the aforementioned one instance in Budongo Forest.

Unfortunately, when Suzuki first saw this incident, the male cannibal was already holding the very young infant victim, so what led to this situation is unknown. Indeed, this is a problem with reports on infanticide among the chimpanzees: observers have yet to see what precipitates such events. In all cases so far reported, their attention has been drawn by the uproar that occurs when an infant has already been seized. But there are a few clues to what may have initiated such behavior.

The cannibal Suzuki saw was the largest male in a large wild group of 80–90 chimpanzees which had gathered at an area he dubbed 'the Picnic Site' (Suzuki 1971:31). The gatherings there were of more than one local group (see Table 1 of Suzuki 1971:32). At the time of the cannibalism, the large Budongo group was exceedingly excited, Suzuki (1971:33) reports. There was always an atmosphere of heightened excitement at these large group 'picnics' or 'booming' sessions (carnivals? overlap areas?). We cannot judge to what extent the presence of the observer heightened the usual initial nervous tension, but we can assume that his presence was somewhat disturbing, because in 1967 the Mahale chimpanzees were not yet fully habituated to observation.

The response of the other males to the seizure of the infant is interesting. Suzuki reports that first one, then another male approached and groomed the cannibal. One, after 'elaborately' grooming the cannibal for about three minutes, snatched the wounded, dangling infant and patted it (Suzuki 1971:33). Other males came close, one occasionally touched the still moaning infant on the back with his hand, which suggests (to the writer) that he responded to it as its being an infant, not as meat.

'All at once,' Suzuki (1971:33) reports all of the chimpanzees around the cannibal became restless. An old male, Mzee, began to climb the tree toward the cannibal 'crying loudly, its hair standing on its end' (Suzuki 1971). The other chimpanzees in the area joined in, vocalizing, which 'touched off cries from all other members of the group separated in three sites in the forest' (Suzuki 1971:33–4). On reaching the cannibal, Mzee presented to the killer, who responded by touching him. Mzee began to groom this male, then reached for the infant and 'touched its exposed flesh' with his mouth. Both Mzee and another male kept reaching out and touching or taking hold of the infant, although they did not try, or were not permitted to take it.

In Suzuki's (1971:33) opinion, the cannibal 'seemed' to feed on the infant, although he tore flesh from the still living and faintly crying infant's thigh and groin area and 'mumbled' it with his mouth, rather than chewing normally. Two hours after Suzuki first saw the male with his victim, the cannibal was still carrying the body in his mouth, apparently little dismembered.

In later months, Suzuki twice saw chimpanzees of this group (including Mzee) eating monkey's flesh. Suzuki (1971:36) reports that, when very large groups gather, the whole group gets extremely excited and that all three cases of meat-eating that he observed occurred when large groups were exceedingly excited. In one instance, the males seemed 'greatly irritated' before they began to chase the monkey victim (Suzuki 1971:39). Natural predators are not normally irritated or angry when hunting.

From the 'tense actions' of the apes toward the possessor of meat, Suzuki (1971) thought it obvious that the chimpanzees liked flesh, even though he points out they do not normally hunt animals. 'The motive that drove the apes to hunt animals' is not the mere fact that the prey is within easy reach, but 'it is also related to social excitement caused by the situation of the group' (Suzuki 1975:268).

A number of cases of provisioned chimpanzees killing and feeding on animals, particularly monkeys and infant chimpanzees, have been reported since Suzuki's sightings. In discussing chimpanzee 'hunting' with other scientists, Kortlandt (1975) points out that the very aggressive facial expression and gestures of a Gombe ape filmed killing an animal are those seen when in socially aggressive intercourse, as when an ape batters a tree or destroys an object. This behavior is redirected social aggression, not true hunting, the scientists conclude. The one case of infanticide and cannibalism seen by Suzuki in a wild group also seems to have been a response to negative social excitement.

It may be recalled that the large (80 member) Mahale M group was habituated to the observers' presence by a method which involved intensive heavy feeding at a fixed feeding station for 4 years, ending in 1974. At that date the free unrestricted feeding was replaced by a method of feeding the animals a very small amount of bait, withheld and doled out at intervals, within the forest. The observers report that withholding the bait resulted in frustration responses from the waiting chimpanzees.

This method was still being followed in January 1977, when Norikoshi observed the first recorded instance of infanticide and cannibalism in this group. Briefly, an M subgroup was encountered in the forest, and the first baits were given out. The second-ranking male began threatening the observers by pounding on the ground and throwing a stone (Norikoshi 1982). The observers gave this male a small piece of the cut-up sugar cane and threw another piece in front of the mother of a 2½ month old infant. (This female was an immigrant from the K group 5 years earlier; the infant was conceived and born in the M group). When the mother reached out to pick up the bait, the angry male attacked her briefly. She retreated from the group and moved away alone with her infant.

One week later, on 13 January the observers were attracted by chimpanzees in the forest vocalizing 'fiercely' (Norikoshi 1982:69). When the observers arrived at the scene, a group of six apes was up in a tree and the same (i.e. second-ranking) male was holding a dead infant, apparently the child of the mother mentioned earlier, as she no longer had her child.

The 'top-ranking' male of M group arrived within a minute of the observers and also climbed the tree and began grooming the infantless mother. A little over an hour later the second-ranking male began to feed on the infant's flesh. Another mother then moved to within 1 meter of the cannibal and watched. Her infant left her breast and climbed on to the head of the cannibal as he fed.

Another male moved close to the cannibal and the top-ranking male moved toward the cannibal and his watchers 'as if he were ready to attack them and displayed by shaking the tree' (Norikoshi 1982:69). This caused the killer to drop the infant to the ground. Another chimpanzee snatched it up and ran into the bushes with it. The observers could not see through the underbrush, but some of the other chimpanzees ran into the bushes and a great commotion was heard, vocalizing and pounding of the ground.

To sum up, over the next 24 hours, four of the chimpanzee held, or fed on, the carcass. Then the observers took it from the top-ranking male who then had it, and was grooming it with his foot. They found only the ears, arms, one leg and genitals were missing, presumably eaten. Norikoshi reports that the first-ranking male repeatedly attacked the individuals who were eating the infant, and so got them to leave the scene of the cannibalism. 'It is open to argument whether the behavior of Kajugi [the top-ranking male] aimed at prohibiting cannibalism or at obtaining the flesh .... The result, however, is that his behavior did help in keeping cannibalism from spreading, to say the least' (Norikoshi 1982:73).

Norikoshi (1982) recalls that, when the female cannibal at Gombe, Passion, took another Gombe female's infant, a male had lent his help in giving the infant back to its mother. He concludes that the evidence is that the infant victim 'was not just another normal meat' (Norikoshi 1982:73). Was the infanticide scapegoating by frustrated and angry males? That question too must be left open at this point.

In the same year (1971) as Suzuki's report of the case of infanticide and cannibalism in a wild chimpanzee group in Budongo Forest, a group of five Gombe (Kasakela, camp area) males was seen to attack an encountered lone wild female and her 2–3 year old infant. The mother escaped, but the Gombe males 'killed by eating' the infant chimpanzee, tearing at its flesh while it screamed (Bygott 1972:410). Virtually the same method of killing by eating is reported at Mahale (Suzuki 1971; Takahata 1985).

Bygott's (1972:411) impression is that cannibalism seems to be a recent behavior that is 'relatively unfamiliar' to the Gombe apes, as they poked, bit, punched, flailed the infant's corpse, or 'spent many minutes gently examining and grooming it' and, in fact, little of the infant was actually eaten. It seems unfamiliar behavior to the Mahale apes – and the wild Budongo apes also, in that Suzuki (1971:33) reports virtually the same ambivalent behavior.

Goodall (1977:279) suggests that these are 'very abnormal' behavior patterns. She points out that if this had been the body of a bushbuck fawn (which would be about the same size as a 2–3 year old chimpanzee), it would most certainly have been quickly and entirely consumed by a group of seven adult chimpanzees. Suzuki (1975:262) reports watching as a male fed 'voraciously' on the body of a blue duiker, a small antelope weighing about 17 kilograms (37 pounds) when full size. In 2½ hours, four chimpanzees consumed the duiker almost entirely.

The victims of the Gombe males were 'stranger' infants, male children of unhabituated migrating females, and all infant killing episodes followed the same pattern. Groups of Gombe males violently attacked 'stranger' females, seized their infants despite the mothers' efforts to protect them, and fed on the carcass in desultory manner. The only female infant victim was badly wounded but not eaten. In a reversal to the protective parental role, several of the males involved gently groomed and carried the severely wounded infant over a period of several hours, until she died (Goodall 1977).

In 12 observed gang attacks on mothers and infants at Gombe, twice the infants were seized, flailed and eaten, twice they were seized, flailed and thrown away. The other eight infants involved were ignored, most continuing to cling to their attacked mothers, although 'the males could have taken the infants' if they wanted to (Goodall 1986*b*:523). Both Gombe and Mahale observers are convinced that the aggression was directed toward the mothers, not the infants (Goodall 1986*b*).

The strangely equivocal behavior of the Gombe males, who alternated between attack and gentle protective behavior toward two of the three 'stranger' infants killed, punching and grooming the carcass, wounding, then caretaking the female infant who died later, brings to mind one of Barker *et al.*'s (1941) findings from their experiments in frustration-induced behavior. When the child-subjects of the famous experiment were blocked from their goal, both the aggression and a state of 'more or less permanent tension' produced by the frustration affected their behavior in many areas of activity. 'Hitting, kicking, breaking and destroying became generalized and frequent' (Barker, Dembo and Lewin 1943:456). Barker found that normal, motivated behavior and frustration-induced behavior are not exclusive of one another, which perhaps explains the ambivalent behavior of the Gombe and Mahale males. As Maier (1961) points out, in actual (human) cases it was found that the two processes are apparently equally strong and so may function simultaneously. As wild, unhabituated chimpanzees normally retreat from would-be observers, perhaps the simple presence of Suzuki at a good and 'ready' (ripe) source of food acted to block, in the sense of making tense, and so frustrate the wary Budongo apes' from reaching their goal of a relaxed and leisurely feeding session.

A number of hypotheses have been proposed in an attempt to explain the phenomenon of primate infanticide. Among the suggestions are a 'mother-into-female' hypothesis, which proposes that infants sired by males of other groups may be killed so that when the female comes again into estrus, the killer may then mate with her, thus enhancing his reproductive success. Other suggestions include population control, disturbance of the physical environment and social pathology (Hrdy 1979). Because of the strong tendency of the Gombe and Mahale victims to be male infants, an interpretation of infanticide as a strategy of males to attain individual reproductive success through eliminating a future competitor is favored by many observers. However, no 'stranger' female whose infant was known to be killed by Mahale or Gombe males subsequently joined the killers' group (Goodall 1977; Kawanaka 1981).

There is also the puzzle of why Mahale M group males killed four of their own offspring, yet permitted three juveniles born in K group to transfer into M group with their mothers (Takahata 1985). Because of this inconsistency, Takahata (1985) rejects the aforementioned explanations of infanticide in primates. He points out that the Mahale killings are not conceivably a male strategy to ensure their own individual reproductive success either by returning the mothers to a state of estrus so that they may then impregnate them, or is it adaptive to eliminate possible future (reproductive) competitors by killing their own offspring. Such behavior does not enhance the killers' reproductive success, it reduces it. He also argues that suggested causes such as overpopulation and environmental disturbances are not applicable to the intragroup infanticide of the Mahale apes, because the population size has not changed in the last decade and, in fact, the M group has expanded its range in recent years to include that of the now-defunct K group (Hiraiwa-Hasegawa *et al.* 1984: Nishida and Kawanaka 1985).

Nishida and Kawanaka (1985) argue that the infanticide and cannibalism at Mahale in recent years is not pathological in that the cannibal-apes do not indulge in any 'bizarre' treatment of the infants' corpses, such as stroking-flailing-grooming-nudging of the infant's carcass, as occurred at Gombe. Was not the behavior of the Mahale mother bizarre, when she permitted her infant to climb on the head of the 1971 cannibal as he fed on another infant's flesh?

The victimized Mahale mothers had both groomed and been groomed by, and also copulated with, the infant-killing males for several years before their infants were killed – and apparently also afterwards (see Table 4 of Nishida and Kawanaka 1985:281). This whole sequence of behavior seems to me bizarre: that is, extraordinary, and involving striking incongruities, maladaptive and pathological. While the physical environment may have changed little, the social and emotive climate of the provisioned chimpanzee of Mahale seems as changed as that at Gombe, and essentially in the same negative fashion.

Opinion regarding the causes and significance of the infanticide observed in the Gombe and Mahale groups is still divided and speculative. Nishida, Vehara and Ramadhani (1979) suggest that cannibalism is a male reproductive strategy in which infanticide (which changes 'mother' into 'female') is extended to include consumption of the victim. Nishida and Kawanaka (1985:283) argue that the Mahale 'within-group cannibalism' observed in later years may be an extension of 'between-group male cannibalism,' hence essentially a matter of 'inter-male competition.' Ghiglieri concurs. He suggests that attacking outsiders and killing infants is a way for the adult males to maximize their reproductive success through exclusion of males of other groups and their 'nonreproductive' (i.e. male) offspring (Ghiglieri 1984:4). Ghiglieri does not comment on the cannibalism aspect of the infanticide, nor on the fact that nongroup females, potential future mates, were also effectively excluded by the attack of the Gombe males, other than to indicate (Ghiglieri 1979) that he saw no instances of either behavior among the nonaggressive, nonprovisioned chimpanzees of Kibale forest. Nevertheless, Ghiglieri (1988:215) now suggests that chimpanzees are 'naturally occurring social killer apes.'[10]

None of these explanations seems entirely satisfactory. All contain anomalies. There is, after all, nothing more crucial for evolutionary success than the raising to adulthood of the number of infants that a foraging range can support, and both the Gombe and Mahale forests are still fairly bountiful. Teleki's suggestion seems more likely, that social stress may cause chimpanzee behavior to 'spark over' temporarily to infanticide and cannibalism from the standard opportunistic pattern of predation of chimpanzees; and that the explanation of the infant-killing incidents may lie 'in the special social and circumstantial conditions within which an episode unfolds' (Teleki 1975:172). We may be underestimating the stress on the chimpanzees of being closely watched and followed by humans – to the apes surely an inexplicable behavior by an alien species.

Persecution of the vulnerable

When all must fight for rank position in an aggressive society, the more vulnerable animals become low-ranking. These include those handicapped by illness, injury, size or lack of strength, the elderly of both sexes, the females, adolescents and children of the group (Goodall 1975). At Gombe by the mid-1970s Goodall (1975:145–7) found that the

> chimpanzees may become aggressive when they compete with one another for social status, or for a favoured food in short supply; .... They may become aggressive, too, if they don't 'get their own way': ... a male, trying to persuade a female to follow him, may attack her if she attempts to escape. A subordinate chimpanzee, who is threatened or attacked or frustrated by a superior, may also be roused to a display of aggression which he usually directs towards some unfortunate individual, lower-ranking than himself, who happens to be nearby. The sight or sound of chimpanzees from a neighbouring community may trigger off displays of aggression, particularly amongst the higher-ranking males, and occasionally 'strangers' are quite savagely attacked .... In their struggle to obtain higher status, some males will take advantage of the temporary disability of any higher ranking individual.

A great deal of the aggression reported at Gombe seems to be less a struggle for rank position than a redirection of aggression – scapegoating, and such well-established frustration responses – in that many of those attacked are females, children and elderly chimpanzees, non-contestants in the power struggle. According to the reports of the unobtrusive field researchers, in the wild groups females and children are not attacked, nor are elderly individuals, even those physically handicapped by age. Quite the reverse, Kortlandt (1962) was impressed by the strong attraction that one sedentary, very elderly male held for the prime adult males. This elder, whose ability to climb and to display was curtailed by his advanced age, was clearly respected by the younger adults. 'Even the biggest of the senior males sought his company' and deferred to him, Kortlandt (1962:130–1) reports. Reynolds (1965*a*) tells of a similarly positive interaction between a youthful wild female and an exceedingly elderly one, whose movements too were slow and labored. The young female followed the old one continually, and now and then groomed her. As mentioned earlier, Sugiyama (1968) and Albrecht and Dunnett (1971) saw badly crippled apes still moving about as accepted members of groups, several years after they were first sighted. Sugiyama (1968:230) mentions that the crippled animal seen in his study group looked relaxed and 'carefree.'

Goodall (1971*a*) on the other hand, writes of a different situation with Mr McGregor, an elderly Gombe male whose legs became paralyzed as a result of a poliomyelitis epidemic in 1966. His former companions began actively avoiding or displaying aggressively against the crippled male, who laboriously dragged himself toward them. The aggressive Goliath 'actually attacked the stricken old male, who, powerless to flee or defend himself in any way, could only cower down, his face split by a hideous grin of terror, whilst Goliath pounded on his back' (Goodall 1971*a*:217). Although the others became accustomed to Mr McGregor's grotesque movements, he was cut off from important social contacts such as grooming and companionship.

Animals vulnerable for other reasons were repeatedly subjected to unprovoked attacks largely by adult males, although others present including females, often joined in these attacks. 'Such "contagion" of aggressive behavior is not at all uncommon in chimpanzee society,' writes Goodall (1975:147).

Prolonged dependency, delayed social maturation

In group-living primate species, outbreeding is usually facilitated by the movement of individuals between groups. In most species it is males who move between groups, while females tend to remain in their natal group. Recent data (post 1965 and 1968) from Gombe and Mahale studies clearly show that the males of these chimpanzee groups do not migrate, but remain in their natal group for life. The only migrants are female, usually young and nulliparous (Pusey 1979). This finding has given rise to an assumption that the adapted pattern among chimpanzees is one of male retention and female migration.

However, Nishida and Hiraiwa-Hasegawa (1986:169) point out that despite adolescent females' showing 'a strong tendency for migration between communities, there is variation between groups.' According to their evidence, females of the Gombe and Mahale groups seldom migrate out. Only two in 15 years emigrated from the Gombe group, while 'at least' 12 youthful females from outside groups immigrated to this group. In the Mahale group, between 1979 and 1983 only three natal females are presumed to have emigrated, and seven female immigrants from other groups (other than the K group) have joined the habituated group. Most youthful females remained mated and gave birth within their natal groups, Nishida and Hiraiwa-Hasegawa (1986) submit.

At Gombe, daily attendance records of visits of chimpanzees to the camp feeding area have been kept since 1963. By drawing on these data and her own observations, Pusey (1979) is able to show that in the mid-1970s, the average age at which young females from unhabituated groups visited or joined the Gombe community was considerably younger than the age at which those growing up in the group left it for the first time. Many incoming females were early adolescents, whereas 'few if any of the females growing up in the habituated community left before developing full estrous swellings; some did not leave until 2 or 3 years after this' (Pusey 1979:477).

This led to what Pusey feels is an unusually high degree of inbreeding. Pusey suspects that the artificial feeding may have caused this inbreeding, and that it has caused a serious general delay in the normal social maturation of the youthful apes. She suggests that the apparent reluctance of the youthful Gombe females to leave their natal group may have come about because, for several years, almost all of the habituated group visited the camp feeding area almost every day, and that consequently the youthful females had unusually little contact with adjacent groups, and far more than is characteristic with members of their own natal group. Pusey (1979:477) suggests that for this reason the females of the Gombe group may have formed 'atypical strong social bonds' with members of their own group. On the other hand, the recent reports of Gombe mothers tending to live alone with their offspring seem to contradict Pusey's suggestion.

One would suspect that the imbalance of overcontact with their own group and undercontact with others should surely have had some effect on the bonding and movement of Gombe males also. It may be that the ambivalent, sometimes protecting sometimes threatening socioemotive atmosphere in which they were raised might retard the social maturity and unnaturally prolong the dependency period of youthful apes of both sexes. It may be that a reluctance to venture forth, based on emotional insecurity, rather than an attraction to stay, underlies the permanent retention of males and the delayed, infrequent transfer of Gombe females beyond the age of independence characteristic of wild chimpanzees. We shall come back to this point.

There is evidence which suggests that the physically maturing Gombe young begin at a very early age to fall behind the wild young in the level of social maturity typical of chimpanzees of their age. The earliest manifestations of a prolonged period of dependency can be found in data on the Gombe mother–infant relationship during the weaning period, which is 'a trying business indeed' (Goodall 1971*a*:161). Before she began her study, Goodall was aware that mothering females under the stresses of captivity usually aggressively initiate weaning and independent locomotion when their infants are between 2 and 3 years of age. Accordingly, she was surprised by the lack of aggression during weaning and the 'rarity of pain-eliciting signals between mothers and infants generally,' in the early (pre-1965) years at Gombe (1968*a*:252). At that time, Goodall could detect no action towards weaning, or rejection of the infants by their mothers. Nutritional independence 'appeared to be initiated by the infant itself' when ready (Goodall 1965*a*:459). Goodall was unable to determine the exact cause which terminated suckling behavior in individual infants but, based on observation from 1960 to 1964, she suggests that 'in the wild chimpanzee mothers do not appear to play an active role in weaning behaviour' (Goodall 1967*a*:302). Furthermore, the records kept by Goodall from 1960 to 1964 indicate that chimpanzee infants are usually entirely nutritionally independent by the age of 3–4 years, though they continue a social dependence on the mother for several more years (Goodall 1967*a*:290; 1968*a*:302). Most infants at Gombe continued to suckle occasionally, for very short intervals, beyond the age of nutritional independence. Goodall (1967*a*:302) refers to these bouts as 'reassurance' suckling, as they tended to occur when the infant was alarmed or hurt.

Flo's daughter Fifi was born before Goodall began her studies at Gombe in 1960. On the basis of a scale of physical and behavioral characteristics thought to typify the various life stages of chimpanzees, Goodall (1967*a*) judged Fifi to be about 3½ years of age when she and her mother first began visiting the camp in 1963, to feed on bananas. Later, Goodall (1968*a*) revised her estimate of the ages of infants born before 1960, upward by 1 year.[11] In 1962–3, 4½ (or 3½) year old Fifi still suckled 'for a few minutes every two or three hours, and jumped occasionally onto Flo's back, particularly if she was nervous or startled' (Goodall 1971*a*:87–8). Apparently the occasional nursing was reassurance, not nutritional, suckling. On the day in 1963 when Flo came into estrus for the first time since Fifi's birth, Fifi was seen to put her mouth to Flo's nipples for a few seconds, then to give up and sit close to her mother, 'whimpering softly' (Goodall 1967*a*:313). She repeated this process several times over the next few days and then stopped trying to suckle. 'Flo was not seen to threaten or push away her child' at any time during this process (Goodall 1967*a*). A few years later Goodall (1968*a*:229) retracts her earlier statement that the 'mothers played no active part in the weaning process,' on the basis that there is 'recent evidence' which shows that mothers do 'in fact initiate the weaning process' through a slow and gentle process which sometimes extends over a period of 2 years.

Flo's next infant, Flint, was born in 1964. In 1966, when Flint was only 2 years and 4 months old, Flo began sometimes to prevent him from suckling. She kept her arms across her breasts when he tried 'to push in and suckle' (Goodall 1968*a*:230). She was never seen to prevent Fifi from having access to her breasts, 5 years earlier.

Even in 1968, Goodall (1968*a*:252) writes that 'not only is there no apparent rejection during weaning, but the child is seldom physically punished by its mother,' as are captive infants. However, while Flo began by gently resisting and distracting Flint from nursing, by 1968 (when he was 4 years old) Flo was observed using tactics such as holding him away from her chest, twisting her body away, walking away while he was suckling, crossing her arms over her nipples, and so on. Flint responded to his mother's attempts to wean him by constantly soliciting attention from Flo. When she rejected his demands, he often flew into 'wild tantrums, screaming, hitting the ground' – even on occasion attacking his mother (Goodall 1973:3). Flo often responded to this behavior by biting or slapping the child. Flint would then respond with more direct violence, 'and they would roll about biting and hitting each other' (Clark 1977:248).

After these conflicts over weaning, Flint would return to his mother and she would groom, hug or play with him, but she would not permit him to suckle. Throughout her publications, whenever Flo is mentioned, Goodall emphasizes that this old female was an outstandingly confident, competent and solicitous mother. Nevertheless, her son Flint became 'a most abnormal' chimpanzee (Goodall 1971*a*:228). Goodall mentions that she could not help empathizing with the frail-looking old female, in the final months of her next pregnancy, when she had to carry not only the unborn infant but also the half-grown 6 year old Flint, 'perched ridiculously' on her back (Goodall 1971*a*:228, and see Figure 6).

Both Goodall (1971*a*) and Clark (1977), who studied the process of weaning of six Gombe infants in 1968–9, emphasize that Flo was the only mother observed to use aggressive tactics in weaning her child, and that Flo's and Flint's was the most extreme case of parent–offspring conflict over weaning seen at Gombe. But other mothers were seen to use some of Flo's tactics, less aggressively, in resisting their infants' attempts to suckle – some when the infant was only 2

Figure 6. Prolonged dependency. Flint still occasionally rode on Flo's back, when he was 8½ years old. (Adapted from Goodall 1975:143) by John Power).

years of age (Clark 1977; Goodall 1971*a*; Nicolson 1977). And, while these females only occasionally rejected their 2 year olds, the rejections intensified and increased in number when the mother's estrous cycle recommenced, usually when the child was about 4 years of age (Goodall 1971*a*; Clark 1977).

Weaning is the first serious break in the bond with the mother, Clark (1977:235) suggests, and having a lengthy (2 to 4 year) period of gradual weaning 'does not completely soften the blow to the infant.' As post-1965 weaning progressed, many Gombe infants showed signs of depression, some for a few weeks, others for a year or more. The weaned infants often became listless and apathetic and sat huddled for long periods. Play and appetite decreased and they sometimes reverted to earlier forms of infantile behavior, such as persistent clinging to their mothers (Goodall 1973; Clark 1977). Until he died at age 8½ years, Flint remained unusually dependent and demanding, constantly 'pestering' and pushing at his mother, almost always sleeping with her, even occasionally riding on her back (Goodall (1971*a*, 1975).[12]

Early weaned !Kung infants react very similarly. Traditionally, women of foraging societies nurse their children for several years, on demand, throughout the day or night (Lee 1979). Although the exact mechanics are not yet clearly understood, it is generally accepted that the long period of vigorous nursing tends to suppress ovulation in women, and probably accounts for the long (3–5 year) birth intervals among the !Kung and other nomadic (foraging) women (Lee 1979). It is efficient that a mother – particularly a nomadic mother – does not produce a second infant until the first one is fully nutritionally, physically and largely socially independent.

Since the !Kung have been forced by government intervention to forsake their foraging life for wage labor or restriction to a native reserve, the birth interval of the women has shortened. Births now tend to be 20 to 23 months apart (Lee 1979). This is 'imposing hardships on !Kung mothers and children alike' (Lee 1979:332).

For the children who are now prematurely weaned at 1.5 to 2 years because their mother is pregnant, instead of 36 months as in the earlier tradition, the trauma of separation is severe, Lee (1979:330) submits. 'Their misery begins at weaning and continues to the birth of the sibling . . . and beyond' (Lee 1979:332). Lee (1979:332) writes of the !Kung mothers having to deal with 'the constant intrusions of an angry, sullen 2 year old.' There is a new and high degree of stress in 'the emotional economy' of the group (Lee 1979:330–2). In Lee's (1979:330) opinion, 'something is out of kilter.'

Studies show that among the symptoms of emotional disturbance in human children are anxious clinging, 'demands excessive or over-intense for the age and situation' or aloof and defiant independence

(Bowlby 1973:24). Many post-1965 Gombe infant chimpanzees manifest such impaired behaviors. Unlike the infants observed by Goodall between 1960 and 1964 (who were usually self-weaned by approximately 3, or just over, years), subsequent offspring of this group who were infants in 1968–9, suckled usually 'until at least 4, more often 5 years of age' (Clark 1977:235). By the age of 4 years a chimpanzee infant is probably not nutritionally dependent on the mother's milk, Clark (1977:246) indicates, but 'it is still psychologically dependent.' Goodall (1975:135) reports that the Gombe young are seldom weaned completely until the fifth or even the sixth year, which is almost double the length of the suckling period of infants in this group, prior to 1965.

The wild juvenile apes of approximately 4–6 years that Reynolds (1965*a*) and Nishida (1968) often saw roaming confidently alone in the forest do not seem to share this psychological dependence. In recent years, if a Gombe child of 6 or 7 years becomes separated from its mother, it begins to whimper and then to scream 'in obvious distress' (Goodall 1973:3–4). Hamburg, Hamburg and Barchas (1975) report that even young adolescents of perhaps 8 years of age often react to their mother's absence by whimpering or crying out in distress. But 'perhaps it is only after a series of such accidental separations' that the young ape initiates 'a brief bout of independent travel', Goodall (1973:3–4) reasons.

Wild infants are remarkably quiet. According to Kortlandt (1962:231) 'they never whined or whimpered'. Even when startled the wild chimpanzee infant usually throws its hands up in the air, without making a sound. (One might expect that a tendency to be quiet rather than noisy in infancy would be selected for, in that noisy infants reveal their presence to potential predators).

Despite the fact that when the Gombe mothers attempted to wean their infants, a lengthy conflict ensued, which resulted in distress for both mothers and their offspring, both Clark (1977) and Goodall (1986*b*) emphasize that the chimpanzee infants still constantly receive a number of 'pleasant' social stimulations from their mothers, such as caresses, affectionate play and grooming. Nevertheless, Gombe infants whimper more or less constantly during the final months of weaning, when the mothers become insistent, Clark (1977) reports. Whimpering usually occurs 'in contexts similar to those which may elicit crying in a human' (Goodall 1968*a*:310). Infants were observed exhibiting distress (by whimpering, screaming, tantrums, rocking, etc.) in the following situations: feeding with mother, grooming with mother, traveling off mother, evening nesting, rough play with others, fear (of sudden noise, snake, etc.), suckling, sex (mother or others copulating), and social excitement and/or aggression, and so on (see Table 7 of Clark 1977:251). This list does not seem to leave many facets of infant life as a source of occasion for adult or child enjoyment.

By 1986 Goodall writes of maternal aggression toward their own infants as normal. In opposition to Bowlby's views on how a child attains the healthy mixture of confidence and normal dependence that a well-balanced adult requires, Goodall (1986*b*:354) suggests that 'a most important function of maternal aggression is to promote independence. A three-year-old infant . . . [is suckled, carried] but during the fourth and fifth years maternal intolerance increases and the child is gradually forced to walk by himself and to feed at a greater distance from her.' (In her 1963*b* and 1965*a* publications, Goodall reports that the gently reared Gombe infants first began voluntarily to venture away from their mothers, to play, at from about 4 months to 1½ years).

Rogers and Davenport (1970:367) found that distress in an infant chimpanzee always produces agitation in the mother. Even totally inadequate captive mothers, who rejected their newborn completely, become visibly agitated by the cries of the neglected infant. Accordingly, there is probably a new tension between the distressed Gombe mothers and their young offspring, a two-way reverberating conflict situation which is in itself stress-accelerating.

It would not do to leave the impression that weaning alone is the cause of infant and mother distress, of infant insecurity, or of a delay in the social maturation (independence) of the post-1965 Gombe group. These young apes are growing up in a disturbed group in which the behavior of others around and toward them is highly and unpredictably changeable. The Gombe young are alternately cherished and protected, or threatened and attacked, by older, larger members of the group. They face a difficult and often dangerous life. But it does seem possible that the sudden, unprovoked attacks from the normally protective males are one reason that the stressed females seem to be trying to curtail their parental investment in their offspring (in order to free, and so protect, themselves) in a social environment in which the infants are more than normally in need of reassurance, comfort and protection from their mothers. There is not this mother–child conflict over weaning in the relaxed wild groups. We would be unsurprised if human children raised under similar threatening conditions were emotionally insecure and needed more reassurance from and contact with their mothers for a longer period of time than the unthreatened, secure children. Most of us would also consider it only common sense to anticipate that a human child's feeling 'ready' (i.e. confident enough, not rebellious enough) to leave the protection of its parents is likely to be earlier if its experience has been with a familiar, nonthreatening 'world', than if it perceives the 'world' as threatening. It is reasonable to assume that the chimpanzee child's response would be similar. Indeed, when the Gombe publications are read in chronological order, they supply evidence of an advancing age for social and psychological independence in the young apes. As we know, in the early years some infants were largely self-weaned between 3 and 4 years of age. Before 1965, Goodall (1968*a*:243) found that juvenile males start to leave their mothers for several days at a time when they are 5–6 years of age and that, by the time a male reaches puberty (at about 7 years of age), 'he accompanies his mother less and less.'[13]

On the basis of core data on attendance at the feeding area, collected by over twenty observers between 1963 and 1970, and his own 1970–2 observations, Wrangham estimates (averages?) the age at which the Gombe young began to travel independently of their mother at 6–9 years of age (Wrangham 1974:83–4). On the basis of her own 756 hours of observation on the Gombe apes in the years 1968–9 and the available banked data, Clark (1977:235) suggests that the young chimpanzee 'remains with the mother for 9–12 years, often spending most of its time with her as its only companion.' The pooled data (from 1963 to 1970) on which Wrangham bases his estimate of 6–9 years of dependence will include Goodall's pre-1965 estimate of independence at 5–6 years, and also Clark's higher (1968–9) estimate of 9–12 years. Apparently the age by which young Gombe apes become socially and psychologically independent has rapidly increased in recent years. By the mid 1970s, Goodall (1975:137) reports that the period of adolescence – postulated in 1963 as lasting from puberty (beginning at about 7, ending at around 10–11 years of age) – begins when the young ape is about 9 years of age and lasts about 4 years in the female, slightly longer in the male, to about 13–14 years of age. Although there is a great deal of individual variation in the attitude of youthful apes toward independence, these chronological reports record a delay in social maturity and a prolongation of the period of child-like dependency among the youthful Gombe apes, which may be in large part due to the traumatic 'emotional economy' of the Gombe chimpanzee group.

Migration among wild chimpanzees

To return to the discussion of migratory patterns of chimpanzees, as mentioned, on the basis of the recent behavior of the Gombe and Mahale apes, it is assumed that chimpanzees as a species exhibit a pattern of female migration from, and male retention in, their natal groups. On the other hand, the evidence from the naturalistic studies of chimpanzees is that (as Turnbull (1968*a*) reports of human foraging groups), there is a free and constant flow of individuals of both sexes back and forth, from group to group.

By the end of 1962, Goodall (1965*a*) confidently recognizes all members of the local chimpanzee groups who fed at the Gombe feeding station. At the time she often saw unknown mature chimpanzees moving about with known apes. Goodall (1965*a*) surmises that they were visiting temporarily from neighboring groups. Attendance records of chimpanzees visiting the feeding station were not kept until 1963, but Goodall's data from 1960 to 1962 'suggest that visiting males outnumber visiting females by three to one' (Goodall 1965*a*:434). She reports 'at least six' 'stranger' males and only two female visitors (Goodall 1965*a*:434). The possible role played by visiting males in breeding cannot be overlooked, Goodall points out.

The Gombe attendance records show that between 1963 and 1965 there was no significant difference in the number of males and females who arrived at the camp feeding area for the first time, a total of 20 independent males and 19 females. Were these newcomers visitors or immigrants to the group – or simply very wary members who had to this date avoided the humans' camp? There is no way of our knowing. Izawa (1970) reports that youthful individuals of both sexes left their natal group from time to time, to move about on their own. Whether they temporarily join and mate in other groups is unknown to Izawa; however, their leaving raises the possibility that wild males and females mate outside their natal group.

Gombe census data (begun in 1963) indicates that 6 males and 11 females immigrated into the Gombe group in 1964–5 (Teleki *et al.* 1976:562).[14] According to Pusey, with the exception of two juveniles, one of whom arrived with his mother, and the other with a young nulliparous female thought to be his sister, 'no males joined or permanently left the habituated community between 1965 and 1972 when the split [into two separate communities] occurred' (Pusey 1979:469).

Although there is a great deal of individual variation in the attitude of youthful apes toward independence, chronologically, these reports also strongly suggest a delay in social maturity, and a prolongation of the period of child-like dependency among the youthful Gombe apes, which, as suggested, may in large part be due to the negative social atmosphere of the provisioned chimpanzee group.

A wild chimpanzee group at Bossou, Guinea, which was habituated to the presence of observers through familiarity, not feeding, was studied by Sugiyama for a total of 340 days between 1975 and 1983. He found that all of the youthful apes in the studied group (four males and two females) disappeared from their natal group during adolescence (Sugiyama 1984). While Sugiyama could not establish reasons for the disappearance of the youthful animals, there were no predators in the area, they were not hunted by humans, and the adult males were not aggressive toward the young. Hence, he thinks it likely that all or most of these young apes emigrated from their natal group. In support of this possibility, recently Sugiyama (1989) is able to confirm that strange males migrate into the Bossou group. When mature adults who disappeared are included in the total of probable emigrants, far more males (seven) left this group than did females (two), which is the usual pattern in most primate species. In view of the mobile, far-ranging role-behavior of males and youthful apes of both sexes, and the more sedentary role behavior of child-rearing adult females, in numerical terms such a ratio of males to females leaving the natal groups, perhaps traveling long distances and joining other groups temporarily or permanently, might be expected. All of the youthful and adult males, but only the few nulliparous, and the even fewer sterile or nonchildrearing adult females are physically free of childrearing, and so inclined to join a mobile group.

Sugiyama's (1984) data demonstrate that migration between groups is beneficial in terms of balancing population numbers. He points out that, during a 6½ year period from 1976 to 1983, the number of chimpanzees (nine) disappearing (probably emigrating) from the Bossou group offsets an increase in population due to births (eight). Indeed, records show that, numerically, the Bossou population has been almost stable since 1967 (see Tables 1 and 2 of Sugiyama 1984:392–3).

Sugiyama and Koman (1979) strongly suspect that the pattern of male retention and female migration reported among both the Gombe and Mahale chimpanzees is a social change, resulting from the artificial feeding. The chronology of the Gombe data on female and male movement between groups suggests that male retention is a change that took place at the time the tension-inducing closed-box feeding was begun. Evolution of a pattern of youthful chimpanzees of both sexes migrating to other groups ensures genetic spread and probably both results from, and facilitates, friendly intragroup relations. Such a solution to population control is far more socially advanced, and more adaptive, than a solution such as infanticide.

Furthermore, although the assumption of a pattern of chimpanzee females migrating between groups and males remaining within their natal group is based on the post-provisioning behavior of the Gombe and Mahale apes, it seems that, while youthful wild females immigrate to these groups, few Gombe and Mahale females actually emigrate to other groups. Nishida and Hiraiwa-Hasegawa (1986:169) report that, during 15 post-1965 years at Gombe,

> only 2 females emigrated from the main study community, while at least 4 natal females remained and gave birth (and 12 females immigrated) (Goodall 1983). Similarly, in M-group between 1979 and 1983 only 3 natal females were presumed to have emigrated, while 2 remained and gave birth (and 7 immigrated from communities other than K-group).

(All of the sexually cycling females of the Mahale K group immigrated to the M group between 1979 and 1982, but this is an 'unusual case,' these scientists continue, due to the disappearance of the adult males of K group, at this time (Nishida and Hiraiwa-Hasegawa 1986)).

It is reasonable to expect that youthful wild females, who have not had the traumatic childhood experience of the Gombe and Mahale apes, would tend earlier to be socially mature and self-directed, more confident – more willing to venture forth on their own – than the socially retarded provisioned Gombe and Mahale animals.

Because the current assumption is that normally chimpanzee males do not migrate, we must be cognizant of the aggressive reception that the Gombe and Mahale males give to incoming 'strangers.' We do not know how many would-be male immigrants, encountering the hostile, patrolling Gombe males in the forest, turn back.

Furthermore, Pusey and Packer (1986:257) suggest that 'sexual attraction to animals in other groups appears to be a common proximate cause of transfer' or migration. The visits or transfers of female chimpanzees to other groups usually take place during their estrous periods, they indicate.

There is no distinct physiological period when it is reproductively advantageous for the males to change groups, but males who are attracted to an estrous female in a neighboring group may easily transfer to that group, or join her, during carnival reunions. When carnivals between local chimpanzee groups ceased at Gombe and Mahale, the males no longer had this natural, inconspicuous opportunity to transfer to another group.

We should consider whether the fierce territoriality of the Gombe males, which ended the important large society carnival gatherings, and the 'fear' and 'hatred' with which they now view 'strangers' Goodall 1990:102, 209) have not also obliterated the normal fluidity of migration of males.

Apprentice into adolescent

Anthropologists usually consider a stage of adolescence, a time between puberty and the attainment of adulthood, to be uniquely human. In societies which have such a period, adolescence involves a varying amount of exclusion. The term implies a period of having outgrown the social status of a child, while not yet being accorded the privileges and rank of adulthood. It is long and difficult in some societies, and is almost missing in others. It is entirely missing in foraging societies, which recognize instead a more positive, apprentice-like period of youth.

Apprenticeship, which may be defined as being a period of specialized teaching or training of novices by more skilled or experienced individuals, has different connotations. It implies a considered degree of appraisal and judgment by adult members of the group, on which full acceptance as a peer depends. The emphasis is on an earning of, rather than exclusion from, acceptance as a fellow-adult. Both these periods, apprenticeship and adolescence, are products of cultural definition, indicative of social, rather than biological, maturation.

Much has been written, in recent years, of the difficult adolescent period of the youthful Gombe apes. Kortlandt does not postulate a period of adolescence among the wild chimpanzees. Rather, on the basis of his experience, he classifies the young wild apes as children up to the beginning of puberty at 8 or 9 years (Kortlandt 1962:130). It is Kortlandt's impression that after that they are, in most respects, accepted as young adults.

In one of her earliest publications Goodall (1963*b*:289), like Kortlandt, reports that, after the chimpanzee child attains puberty, during the next 3 or 4 years of development 'it gradually takes place in adult society.' Goodall refers to this period among the Gombe chimpanzees as 'adolescence,' but it is an age of apprenticeship, rather than of adolescence in wild chimpanzee groups.[15] There is no suggestion or evidence of exclusion of the youthful apes by aggressive means in Goodall's writings based on data collected before the frustrating method of feeding was introduced in 1965. In fact, in the very early years (1961–2), Goodall (1963*b*:289) was impressed by the 'particularly harmonious' relations between mature and youthful males during the latter's 2 to 3 year transitional period to adult rank: 'they do not even fight over females!' Adult males did not control mating. Goodall reports that youthful males copulated with sexually receptive females along with older males, and there was no sign of jealousy or antagonism between them. She also noted that few of the youthful apes seemed to need the occasional adult correction or discipline that chimpanzee children sometimes bring on themselves as they grow older and, socially, more is expected of them. Other than these points, Goodall (1963*b*) reports no other distinguishable behavior typical of the youthful period of a wild chimpanzee's lifespan.

Most of the researchers who used naturalistic methods report a relaxed and amiable acceptance by adult males, of youthful members of both sexes, as traveling companions (Azuma and Toyoshima 1962; Reynolds and Reynolds 1965; Albrecht and Dunnett 1971; Ghiglieri 1979). Ghiglieri (1984) reports that even older juvenile apes often travel with the wild adult males of Kibale Forest. Sugiyama (1972:150) suggests that the supportive social environment of the wild chimpanzee groups 'allows a youngster to move freely, without restriction.' In Budongo Forest, older juveniles often leave their mothers to travel with familiar adults with whom they feel comfortable and assured. The independence, or dependence, of youthful chimpanzees appears to be (as Sugiyama surmises) a matter of individual inclination and readiness in the sense of social maturity, confidence and ability to take on the adult roles.

There is considerable evidence which indicates that in wild chimpanzee groups youthful males and females must achieve adult rank in the same positive manner as do youthful human foragers. They must impress on the group generally that they are socially mature young adults, ready and able to take on adult roles. The methods by which they do this are more fully explored in Part 6 (Achievement of adult rank). The reason for bringing a general impression of the period of youth among wild chimpanzees into the discussion in this section is to emphasize the extent of social change that has come about in yet another aspect of the social life of the Gombe chimpanzees.

In the 1970s, researchers from Stanford University investigated the psychobiology of adolescent anger and depression among chimpanzees, some in artificial (caged) colonies, but most 'in the natural habitat,' i.e. at the Gombe Research Center (Hamburg *et al.* 1975:242). These researchers report that the youthful Gombe apes have a surprisingly long, difficult, 'emotionally turbulent' period of adolescence during which the youthful males particularly are socially and spatially peripheral to the group (Hamburg *et al.* 1975:244). The turbulence and peripheralization of male adolescents takes many forms, yet with a constant theme which is, Hamburg *et al.* (1975:245–6) explain, 'the upsurge of aggressive behavior and the struggle for enhancement of dominance status.' Hamburg and his colleagues view this period of adolescent struggle which the youthful Gombe chimpanzees endure as a period when skills and behaviors learned in childhood can be practiced in the adult context, in order to prepare the youthful animal to enter social maturity. Goodall (1975:137) also views the adolescent period of the Gombe young as 'a further extension of the period of immaturity,' in which the youthful apes perfect social patterns 'such as the spectacular charging display in the male and care of infants in the female.'

During this adolescent period, the male has a growth spurt, after which he tends to become more aggressive, particularly toward females. At some point in his development, Goodall (1975) continues, when he is confident enough, large enough and aggressive enough, the adolescent must assert himself aggressively in order to rise in the hierarchy. Having begun by dominating smaller apes and child-protecting females, he now uses display actions 'in a number of status conflicts' with 'other males of similar age and social rank as himself with whom he associates quite frequently' (Goodall 1973:7). According to the pioneer observer, these display actions are usually bluff, but may involve actual physical attacks.

During this period, the adolescent Gombe male learns that he must be very cautious around the mature males, to avoid arousing aggression. Some are quick to threaten him for behavior that they tolerated when he was a juvenile. 'Nevertheless, despite the fact that he may become increasingly fearful of these older males, he often seems to deliberately choose to associate with them' (Goodall 1973:7). Both Goodall (1975) and Nishida (1979) remark on the attraction adult males have for youthful males, despite their often being attacked by the older males if they try to associate with them. At Gombe, adolescents tend to sit as close to groups containing adult males as is tolerated, but, if one of the older males suddenly moves, an adolescent may jump away or whimper (Hamburg *et al*. 1975). The excluded or barely tolerated Gombe adolescent often either travels alone or returns to the company of his mother, particularly if he has been a victim of the males' aggression. Hayaki (1985), on the other hand, reports often seeing adult Mahale males waiting for youthful males to follow them.

When youthful wild male chimpanzees are physically peripheral to a mixed group, they are not necessarily socially excluded. It may be that they are distancing themselves from their mothers, and from the children generally, in order to signal that they identify with, or wish to be identified with, the adult males. Although it is surmise, it seems likely that, when the youthful chimpanzees are not competitors but apprentices, their being apart is probably based on self-direction. By thus demonstrating independence, which is a quality signifying adulthood (and a property of the social system), the youthful apes display their desire to leave childhood behind and their readiness and ability to be considered of adult rank.

In contrast to her report that relationships between then-wild mature and adolescent males were entirely harmonious, (Goodall 1963*b*), with no sign of antagonism between them even when the youthful males copulated with estrous females, 14 years later, Goodall (1975:138) reports that, when the adolescent Gombe male displays, he does not dare do so near the adult males, but 'somewhere off to the side, in the bushes.' He may be permitted by the older males to copulate, Goodall continues, but only after they have done so: usually he mates while screened by a tree, out of sight of his superiors.

Gombe adolescents are excluded from sharing the banana bait. Goodall (1971*a*:174–5) comments that there were times when, as a young male sits apart, watching a 'high-ranking' adult male feeding on bananas, the tension appears to build up until 'the adolescent has to give vent to his frustration in a charging display' and he rushes off crashing through the undergrowth. Unfortunately it seems 'that adult chimpanzees are often irritated by a . . . commotion from a youngster', so the adolescent 'may well be chased and even attacked for his ill-advised display,' Goodall continues.

By the time a young Gombe ape is about 13–15 years old and of full adult size, he begins to threaten and occasionally to attack adult males who are low in the ranked hierarchy. When a youthful male is able to subdue even one of the adult males consistently, he can be considered as part of the adult hierarchy (albeit of low rank) and so socially mature, Goodall (1968*a*, 1975) suggests. After gaining this acceptance, the young male charges into a camp along with the adult males, and takes a turn 'at courting attractive females during the frenzy of excitement . . . instead of waiting until things were calm and their elders less worked up' (Goodall 1971*a*:178). By means of aggressive competition, the youthful Gombe male now forces his acceptance as an adult, with a position in the aggression-based rank order.

A female chimpanzee reaches puberty at about 7 years of age. At this age she shows small irregularly occurring swellings of her sex skin, although she will not menstruate or attract the sexual solicitations of the mature males for about another 2 years.

What of the experience of the youthful Gombe female in recent years? Goodall (1975:138) suggests that, in general, the female adolescent does not encounter such a difficult proving time as the male, because 'she is less concerned about establishing her position in a hierarchy' and is 'content to stay with her mother.' As the mother usually has another infant 'she can learn all she needs to know about maternal behavior without going further afield'. As adult chimpanzees of both sexes must be self-sufficient for food (particularly in the dry season when food is scarce and the summoned larger group gatherings at a food source cease), and because a female should mate outside of the natal group, there is a great deal more a youthful female must learn besides maternal behavior. Foraging efficiency increases with experience, and the youthful chimpanzees of both sexes must learn about their geophysical environment and its resources, and establish friendly social relations with adjoining groups. The most efficient way of learning to do both of these things is by the example, and imitation of experienced and skilled adults, through traveling for a time as members of the far-ranging mobile groups.

The evidence is that the youthful Gombe female too experiences an adolescent period of being excluded. She, too, 'must be cautious in her dealings with her social superiors' – in her case, not only the adult males but also the adult females and adolescent males as well (Goodall 1971*a*:178). Goodall (1975:147) reports two cases in recent years in which 'older females, for periods of several months "persecuted" 2 young adolescent females,' chasing them off repeatedly, until they began to scream and hurry away whenever they saw the aggressors approaching. Even 'the precocious juvenile offspring of high-ranking mothers' sometimes harass adolescent females (Goodall 1971*a*:178).

In post-1965 Gombe, young females tend to remain with their mothers until they become sexually attractive at about 11 years of age, Goodall (1975) reports. As we know, this is much later than is the case with wild females, some of which are traveling independently while still early adolescents (not yet sexually attractive to adult males, hence probably between 7 and 9 years of age (Pusey 1979)).

It is shown in Part 6 that, as among human foragers, in wild chimpanzee society acceptance as an adult is earned through individual social development – demonstration of social maturity and competence. It is not fought for. There is no record of how youthful Gombe females gain acceptance as an adult in their natal group. They are eventually accepted by the males as sexual partners (Pusey 1979). But we do not know if they also gain social acceptance, as full members of adult rank. We do know that Gombe females now sequester themselves, or are excluded, from male-only groups (Wrangham 1979). Each female tends to live, separated both from males and sometimes from other females, with only their dependent offspring for company.

Solidifying alliances

Partner preferences are common among chimpanzees, but the stable, closed alliances formed through a personal relationship with one particular ally (or group of allies), which are typical of many species of primate, are not. The social system of chimpanzees is one in which fluidity of partnership is essential. In order to maintain a high egalitarianism and equal individual autonomy in a system based in part on a leader–follower relationship, the pattern required is one of constantly and easily changing *partners*, *roles* and *statuses*. (The only *ranks* are adult and child. Rank as an *adult*, once achieved, does not change).

Closed alliances are potentially hostile toward each other, and are no more functional in terms of protection against attack than the fluid, shifting charismatic–dependent coalitions. Indeed, the usual coalitions, i.e. temporary alliances, provide the benefits of stable alliances without the disadvantages and dangers involved in more permanent units.

Partner preferences among chimpanzees often develop from long friendships, perhaps based on childhood playmate or sibling relations. They are particularly strong when the added ingredient of complementary personality types is involved. As a collateral of the aggressive direct competition at Gombe, the preferred partner coalitions tended to solidify, to 'freeze' into stable alliances between certain charismatic and dependent individuals. Changing partners became less free and less frequent. Whereas formerly almost any of the charismatic apes might travel or socialize temporarily with any of the more dependent animals attracted, under competitive stress at Gombe the same pairs began to stay together for longer periods, as the mutual defense aspect of the relationship became more urgent. By the 1970s some pairs of adult males formed strong supportive relationships lasting for years (Goodall 1975).

This 'solidifying' – it would be misleading to refer to it as stabilizing – permitted Goodall to identify a number of pairs of chimpanzees who tended to spend much time together. She realized that these fairly constant combinations were usually made up of pairs who were remarkably different in temperament. It is Goodall's (1971*a*) vivid descriptions of the personalities and interactions of these atypically long-lasting coalitions which first permitted my identification of the charismatic–dependent relationship. Among the frustrated and directly competing Gombe apes, however, physical strength and the quality of aggressiveness became important. In the solidifying alliances, qualities of amiability, trust and friendship are no longer the most important bases for a relationship. Compatibility became less important than a need for protection from aggressive members of their own group. It follows that the quality of dependency would change from a (positive) trustful reliance on, to a (negative) submission to, another (or others) in return for support.

A subjective experience of weakness and insecurity is thought to be connected with this negative form of dependence in humans (Bowlby 1973). It seems reasonable to assume that the submissiveness Gombe observers so often report in not only the young, the females and the timid of both sexes, but also adult males lower in the hierarchy than the current despot, is coupled with a similar psychological experience in the deteriorating social climate.

PART

4

The behavior of wild and provisioned groups: a theoretical analysis

Foraging theory: when, and when not, to defend territory

The degree of exclusive use of a territory varies widely among animal species, from almost total control to sharing significant portions of the area and its resources with outsiders (Myers, Connors and Pitelka 1981). Animals do not usually defend a territory, excluding others, unless the benefits of so doing exceed the costs. In this section, the behavior of the wild chimpanzees, and the provisioned Gombe and Mahale groups, will be considered in terms of foraging theory (costs and net benefits), frustration, and competition theories.

Customarily, the term foraging is used to refer to behavior associated with searching for, obtaining and consuming wild foods. Foraging *strategy* refers to the way in which foraging animals achieve a particular goal. Animals that adopt the most successful strategy forage 'optimally' and are favored by natural selection (Krebs 1981). The goal and criterion of success might be to obtain as much food as possible or to minimize the time spent foraging each day in order to 'bank' time and energy for other activities, such as seeking mates, reproduction, resting, grooming, antipredator behavior and so on. Foraging strategy is often viewed in terms of the tactical 'decisions' that animals make, as to which food items to seek and when, where and how to do so. As Krebs (1981) puts it, in terms of fitness the foremost question is: what tactics should an animal adopt that will be most effective toward reaching its goal?

The primary requirements for a species to survive are food and reproduction. The whole psychology and social structure of a species, its behavior and social needs, is a secondary adaptation to these primary life requirements. The foraging pattern that a species develops is a secondary adaptation, designed to serve the basic needs most efficiently within the particular environmental limits. While the ultimate goal of foraging behavior is to acquire food, highly evolved social animals, such as chimpanzees, have more complex goals. I suggest that an important goal of chimpanzees is to incorporate a beneficial sociality with the basic food search, and that they have developed special tactics to do so which at the same time optimize the chances of survival of the individual and, ultimately, the species.

According to Krebs and Davies (1981:57), optimality theory starts with the assumption that natural selection has designed optimally efficient animals. Ultimately 'efficient' animals are those which are 'good at getting genes into future generations.' Efficiency must include success in somehow ensuring that a next generation reaches the age of reproduction. Success measured in terms of reproduction alone is valid in species whose young are born at a developmental stage of independence – many species of fish, for example – but success among chimpanzees cannot be measured in terms of reproduction alone. Their young have a lengthy period of dependence on adults, especially the mother, for food and protection.

Optimal foraging theory also postulates that natural selection will favor those animals that forage most efficiently. In some circumstances, this efficiency is gained by defending a territory, excluding others aggressively; in others, it is achieved through nonaggressive spacing out attained by the signaling of occupancy in some way. In a summary of current ethological findings regarding when, and when not, an animal should defend a particular territory, Myers *et al.* (1981:136) cite Brown's widely accepted concept of defensibility, which is that 'an animal should defend when the benefits of defense exceed the costs.' These authors write in terms of individuals, but, as they point out, each is part of a larger community and any behavioral adjustments by the individual animal will affect its interactions with other elements within the community. As we know, territorial defense as it appears among the provisioned chimpanzees is not individual, but the action of groups of males.

There are both costs and benefits to defending a territory. Myers *et al.* (1981) suggest four key points which affect whether and to what degree an occupant animal restricts access to an area by defense. (1) Restriction of access is through aggressive defense of the site. Aggressive defense of an area is beneficial if the occupant animal gains an increase in resources (food and mates) by excluding conspecifics.

(2) Territorial defense of a particular site entails an expenditure of, or investment in, time and energy. (3) To justify the investment of time and energy, defense must yield net benefits (benefits minus costs) greater than the animal would obtain through using various forms of spacing behavior, which achieves even dispersion patterns through mutual avoidance without overt aggression. (4) When resource density or availability is significantly reduced by use, and animals are able to control access to resources, defense appears to be the best strategy. However, Myers points out that there are also nonenergetic costs for territoriality. Among the provisioned chimpanzee groups, the social costs for defending their home range territories are high indeed.

Costs and benefits of recent behaviors of the Gombe and Mahale apes

Before we consider the behavior of the Gombe and Mahale chimpanzees in terms of foraging strategy, the reader is reminded of four earlier established points. First, typically wild chimpanzee home ranges overlap fairly extensively with those of neighboring groups, and both are reported to use the overlap area amiably, often at the same time. Second, in typical chimpanzee habitats, sources of wild foods are patchy in their distribution but not scarce. Individual fruiting trees of the same species are scattered and ripen at different times. Third, the chimpanzee diet is omnivorous and eclectic. Fourth, despite the spatial erosion of the natural habitat by humans, there is still more than sufficient wild food in the Gombe area (Goodall 1990). Because the observers do not allude to shortages of wild food in the quite similar Mahale habitat, I assume there is sufficiency in that area also.

Overlapping sections of neighboring home ranges tend to be areas in which certain preferred foods are more abundant than elsewhere (Nishida 1968; Goodall *et al.* 1979). Since two, or several, neighboring chimpanzee groups utilize these areas of choice food as part of their foraging range, the problem of territory usage had to be worked out. It was in such an overlap area that both the Mahale and the Gombe observers chose to set up their permanent feeding stations.

Until 1971, all of the regularly provisioned apes at Gombe mingled so freely that they were thought to belong to the same group. After 1971, it was decided that these were probably two neighboring local groups, and that the Gombe feeding area was in or near to the area where their home ranges overlapped. In 1971, association between these animals gradually decreased, and, by 1972, the groups had separated permanently into two increasingly hostile communities.

Right from the beginning of observation at the permanent feeding station at the Mahale Research Center, the resident chimpanzee groups appeared to solve the problem of sharing some territory by using a strategy of spacing out, by vocalizing as they approached the shared foraging area, followed by retreat or avoidance by one or other group. Of course, the Mahale observers cannot be sure that avoidance was the chimpanzees' solution to sharing this overlapping food source before their arrival. It is Reynolds' (1965*a*) and Ghiglieri's (1984) impressions that the vocalizing seems to attract others, rather than space them out. When two groups meet at a food source, there is probably always a problem of tension which must be resolved. It may have been that a heightened tension, caused by the presence of humans, caused the Mahale group to avoid the other occupant group. Avoidance is undoubtedly a solution which is adaptive in terms of solving the problem of joint tenancy, but it does not facilitate intergroup sociality, or, it appears, interbreeding.

Pioneer researchers Itani and Izawa observed two large groups of wild chimpanzees in the general area of the Mahale Mountains for three months in 1963. Izawa (1970) reports that, although their home ranges overlapped by approximately 20%, the two groups routinely foraged the overlap area at different times. Izawa (1970) reports that on one occasion, Itani saw the vocalizing groups come to within 200–250 meters of each other, without mingling. 'No special tension or excitement on the part of any individual was detected' (Izawa 1970:36). Neither group showed any sign of hostility toward the other.

For 11 months, from October 1965 to January 1966, Kano (1971:239–40) also observed two large groups in the same general area, which he never saw mingle, 'even for a moment' throughout the entire observation period. (It is possible that these are the same wild groups which Itani and Izawa observed in 1963). Kano does not report any signs of aggression between these apes. Still, these reports suggest that wild chimpanzee groups may solve the problem of sharing resources and territories either through the model proposed, of free 'carnival' mingling, or through understood separate nonhostile occupancy. While the groups do not mingle, it may be that lone individuals move between these nonhostile groups, as happens in groups that do mingle. The point is, even if normally separate, being without hostility toward the neighboring local group facilitates solving the problems of joint occupancy (and perhaps interbreeding) without the high energy and social costs of territorial defense.

Established defense of a territory is only possible when it is of a size and in an ecological location which renders control of access possible through aggressive means. Goodall (1965*a*:456) establishes that in the early years at Gombe (1960–2), all mature, individually recognized chimpanzees, with the exception of mothers with infants, 'have a range of at least 20 square miles' – approximately 52 square kilometers. Generally females with young range less extensively, but that depends on the individual animal. Some mothers were seen to range as far as did the males. In 1965, before the group split, there were 29 males from infants to elderly (Teleki *et al.* 1976). Even if all 29 males had been prime adults, defense of a range of such magnitude is impossible.

Wrangham (1979) analyzed the ranging of the Gombe males over a 12 month period in 1972–3. By this date, the Kasakela (camp area) group and the Kahama (split-off) group occupied partially separate community ranges, covering approximately 13 and 10 square kilometers respectively. Wrangham does not report the size of the individual ranges of the Gombe females, other than that they are much smaller. At the time of Wrangham's (1979:483) analysis there were nine adult males in the Kasakela and seven in the Kahama groups. Drawing on Wrangham's data, Riss and Busse (1977:285) report that the mean daily average range of the adult males at the time was 4.9 kilometers. Some males rarely traveled to the periphery of their curtailed range.

By 1981 the Kasakela group's range 'was down to about 9.6 square kilometers – half its 1977 size.' Later, that same year, apparently, with the group down to only five mature males and 18 females and their young, 'the total range shrank to around 6 square kilometers,' scarcely enough to support the group, Goodall (1986*b*:516) submits. Even a range reduced to 6 square kilometers cannot be effectively defended by five males, even though, if they were now somewhat short of food resources (which they are not), there might be selective benefit in defending the territory. (When resources are abundant, defending a territory carries no selective benefit).

Exclusive access to resources and to breeding females is a net benefit, maximizing fitness only if the costs do not outweigh the benefits. We know that, according to recent reports, some Gombe and

Mahale females leave the group to mate, but the males apparently do not (Pusey 1979). Mates for both must come from outside groups if these groups are not to become inbred. Nishida *et al.* (1985) agree that theoretically males should welcome unrelated females, rather than try to kill them. Yet, Gombe males fiercely attack and often seriously wound 'stranger' females, who then retreat from, rather than join, the attackers' group. Yet, as Goodall (1986*b*) points out, these females are not always strangers. She suggests that some of them may have been encountered before, some several times. 'They may even have visited the aggressor males as late adolescents during periods of estrus' (Goodall 1986*b*:523).

Through infanticide, the Mahale males (and perhaps the Gombe males, also) are reducing rather than maximizing their fitness, in that the infants that they killed either carried their genes or were genetically desirable potential mates of the young of their group.

Some primatologists accept killing or driving away encountered outsiders (which at Gombe includes adult males, females and young) as a drastic but effective way of maintaining an exclusive territory. It may be recalled that Ghiglieri (1984:4) hypothesizes that through excluding nongroup males and killing 'stranger' male infants, territorial male chimpanzees maintain exclusive access to their groups' breeding females and the forage resources, thus maximizing their individual and inclusive fitness.[16] Since four of the Mahale infant victims are known to be offspring of the males that killed them, Takahata (1985) rejects the idea that the motivation for this infanticide is the goal of improving the killers' reproductive opportunities or eliminating future competitors. There seem to be no reproductive benefits. Rather than expanding their present and future opportunities for sexual reproduction with genetically desirable mates, the Gombe and Mahale males are reducing them dramatically.

Moreover, there are costs in terms of nutrition. Food resources are not evenly distributed throughout the area. By closing their borders the Gombe apes do retain sole use of resources within their reduced range, but they lose access to foods in other areas. Their diet is less varied. Efficient foraging is learned, and the young of the group are no longer learning the full variety and sources of wild foods available to chimpanzees. All this makes the group more vulnerable in time of food shortage, drought or failure of a local crop. By reducing their range there will be a saving in time and energy expended on foraging, but this saving may be offset by the time and energy spent in patrolling and defending their territory. Because there are no comparable statistics, we do not know.

There are also high social costs to both provisioned groups. Social attitudes are learned. Particularly at Gombe, where the mothering females are alienated even from their peers, the opportunities for the young to learn expected behaviors from observation of interactions and relationships within and between groups are significantly reduced. The presence of adults and peers during the childhood period, when most social learning occurs, is important for survival and later reproductive success in most primate species (Broom 1981:190).

Foraging theory holds that when the food environment of an animal species is patchy, selection would favor increased gregariousness and cooperation (Kurland and Beckerman 1985). As Anderson (1968:154) suggested of human foragers, in circumstances that are to be constantly changing with the seasons, weather conditions, demography, and resource supply, 'what better security is there for the future than to keep a wide range of strategic social ties "warm"?'

By closing their home range territory to outsiders, the provisioned chimpanzees have isolated their groups and reduced the important pattern of fission and fusion from inter- to intragroup level. By doing so, they have lost many social and political benefits (see also Power 1986).

The nonacquisitive, anti-authoritarian, individualized, egalitarian foraging societies are not without problems, but, as Silberbauer (1981:463) points out, maintenance of positive intra- and intergroup relations is 'the supreme value and ultimate rationale for much that [foraging] people do.' Perhaps more appropriately applied to chimpanzee behavior is Leacock and Lee's (1982:9) suggestion that 'the social life of foragers is in good measure the continual prevention or working out of potentially disruptive conflicts.' Indirect competition – the carefully maintained spacing out of chimpanzees feeding in a tree that impressed Ghiglieri (1988), and the pattern of fissioning into single individuals searching for food when food is scarce – acts to prevent direct confrontation over food. But the whole pattern, both fission and fusion, can be understood, in anthropological terms, as a fundamental means to the ends of both avoidance or resolution of conflict, and for maintaining and regularly positively reaffirming multiple enduring social ties both within and between local groups.

The fission aspect of this pattern functions to disperse tensions *before* they build up and threaten the harmonious relationships of the group. When quarrels break out, one or the other disputant leaves the area – exits, rather than flees, until tempers cool. Just the possibility of conflict, or simply a feeling of vague dissatisfaction with the current situation is sufficient to cause human foragers to scatter, often abandoning camp within a few minutes (Woodburn 1972:201–2).

Fission allows individuals to segregate themselves easily from those with whom they might come in conflict, without social or economic penalty, or sacrifice of other vital interests, thus subverting the development of coercive power (Woodburn 1972). It acts to distribute people rapidly in relation to resources, and at the same time functions as a powerful leveling mechanism. An important point, which Woodburn (1972:205) suggests, is that when foragers move they are not only disassociating themselves from some people – but usually associating themselves (fusing) with others. Individuals who exit from a group often visit other local groups for a time.

The fission–fusion pattern functions at both individual and group level. Local groups gather (fuse) periodically in larger society gatherings, which take on an atmosphere of carnival in both chimpanzee and human foraging societies. Silberbauer (1981:462) suggests that the periodic reunions as a larger society overcome the autonomous character of separate individuals and human family (subgroups) because of the advantages of information exchange and 'from the very high value which the . . . [foragers] place on the company of their fellows.'

An extremely important point is that the pattern of constant fission and fusion functions to oppose formation of any corporate system or formation of exclusive groups – even kinship – other than in the forementioned broad sense of social kinship (which it reinforces). A pattern of fission and fusion cannot long coexist with a rigidly defined structure such as an aggression-based dominance hierarchy: each mitigates against the other (Lee and DeVore 1968; Turnbull 1968*a*). Chimpanzees, or humans, cannot be organized into some form of imposed hierarchical rank, headed by a dominant who gains his alpha rank through aggressive conflict with his fellows, while at the same time following a full form of this aggression- and coercion-avoiding pattern. As Silberbauer (1982:33) points out, the openness of groups as social units eventually brings about the defeat of forceful factions because reluctant members would exercise their freedom to move to another group.

The Gombe and Mahale apes have lost much by closing their borders, and curtailing the fission–fusion pattern.

The foraging strategy of wild chimpanzees

The wide ranging which was part of the foraging strategy of the Gombe apes before provisioning must have had some strong beneficial attraction or advantage, since these apes did not need always to range so widely to obtain food. Itani and Suzuki (1967:371) note that wild chimpanzees change groups and move from one area to another long before the food supply is depleted. Human foragers, too, move from place to place 'far more frequently than is strictly necessary if movement is seen simply as a means of providing the best possible access to supplies of food and water . . .' long before shortages have become in any way serious (Woodburn 1968*b*:106). Among human foragers, social reasons for moving seem to be stronger than the obvious ecological ones (Turnbull 1968*a*; Woodburn 1982). It was Itani and Suzuki's (1967) impression that the constant movement of wild chimpanzees around the home range also did not seem to be an adaptation to food shortage. Still, according to foraging theory, there should be a net benefit.

It will be recalled that when a mobile chimpanzee subgroup arrives at an ample source of wild fruit that is ripe, they vocalize, which attracts other roaming mobile individuals and groups and those of the more sedentary groups who choose to respond to the calling. Any individual or group of either sex – of the callers' group or not – may respond to the calls and join what becomes a temporary gathering of a large mixed society often made up of responding subgroups from neighboring populations.

Sugiyama twice watched chimpanzees of neighboring groups meet in an overlap area. They mingled, 'exaggeratedly' ate leaves and fruits, and ran, leapt, drummed and so on, in highly excited but entirely nonaggressive socialization (Sugiyama 1972:155–6). Sugiyama (1972) suggests that friendly social relations between neighboring local groups are maintained by this positive, seemingly ritualized behavior. (Interestingly, and perhaps significantly, Goodall finds that her miming the apes picking and eating wild foods seems to convey to the animals the message that her intentions are not threatening).

Reynolds (1965*a*) points out that not all chimpanzee gatherings lead to the high positive excitement of tumultuous revelry. As indicated in Part 3, after observing several of the larger group gatherings, Reynolds and Reynolds (1965) formed the impression, which they could not confirm, that it was *when subgroups of differing local groups came together* that the gathering took on a carnival atmosphere. I argue in Part 6 that the carnival reunions bring together a larger, constantly fissioning and fusing society of which local groups are units.[17]

At these larger society gatherings, after an initial period in which the apes dash about in excited, nonaggressive display, which acts to relieve tensions (Kortlandt 1962; Sugiyama 1972), the apes settle down for what appears to both Kortlandt and Sugiyama to be less a feeding period than one of intense socialization, which acts to strengthen social relations between individuals and neighboring local (home range) groups.

This behavior is in accordance with foraging theory. In addition to suggesting that increased gregariousness is beneficial when a food environment is patchy, Kurland and Beckerman (1985:85) indicate that search costs for scattered foods can be reduced for all individuals, and individual foraging yields increased, if a number of foragers – whether related, or unrelated, members of the same group or not – cooperate by ranging over different parts of the habitat and exchanging information about located food items. When the more sedentary members of local groups, the mothering females and the elderly of both sexes, respond to the calling and move to join the gathering, there are savings in terms of energy and nutritional benefits for these members. In addition, when the mobile groups move from one good feeding area to another, and sedentary members follow, in a constantly repeating pattern, the whole home range is utilized in the food search; no one area is overexploited.

There are also other important benefits, which Gombe and Mahale apes have sacrificed by becoming aggressively territorial. For example, at chimpanzee carnivals much information is exchanged through observation, such as which females are or are not in estrus, which have new infants, which youthful apes are now being accepted as adult, and so on. Impressions are formed of others' characters and abilities. These reunions provide opportunities to make new friendships and renew old ones, and for the young to learn, by example, appropriate social behaviors and attitudes.

After one to several hours of carnival, subgroups begin to reform, typically with different members (some from other local groups), and the apes drift off at will to resume their separate foraging rounds until they are called together once more, to repeat the socially cohering reunion. When individuals join and move about with a neighboring group, the population is redistributed, and all learn the resources of a much larger range. Through constant encounter under positive conditions, neighboring local groups become familiar and friendly, permitting social interaction and a network of positive relations to extend beyond the local group, to a larger society. This beneficial phenomenon has been severely curtailed by the exclusive territoriality of the Gombe and Mahale chimpanzees.[18] As was emphasized earlier, positive social relations within and between groups are structural and functional necessities in a primate species following a fission–fusion pattern and an immediate-return foraging system, if the loose structure is to hold.

While a great deal of time and energy is invested in carnival gatherings which could be utilized in further food foraging, finding food is not a problem in the natural habitat; and as Teleki (1981:328) points out, chimpanzees can operate at different levels. They are 'quite capable of performing actions that are energetically costly when there are social advantages to be gained.'

In summary, the open, joint use of overlapping home ranges and their resources by the wild chimpanzees seems to be an important structural element of the optional foraging strategy for immediate-return foragers, both human and chimpanzees. It maximizes nutritional and reproductive opportunities while enhancing a high and beneficial sociality, without the costs of territorial defense.

There seems to be no net benefit accrued through the recent behaviors of the Gombe and Mahale chimpanzees. Through closing their territory and consequent heavier feeding in a smaller area, they probably risk degrading (and do limit) their environmental resources. Through this action and an agonistic relationship with neighboring groups, they have reduced their nutritional and reproductive opportunities and decreased their sociality.

It seems that quite without such intent, the observers who provisioned the Gombe and Mahale apes set up a human-made sociopsychological environment involving direct competition and frustrating conditions. This unnatural environment has changed the behavior of the provisioned chimpanzees at the expense of the primary or normal behavior and organization. I suggest that the recent behavior of these chimpanzees can best be understood in terms of frustration and competition theories.

Frustration theory

Frustration is a state known to be shared by humans and animals; the general theory that has been developed is applicable to both animal and human behavior. We cannot tell whether or not chimpanzees experience the same emotional state as humans would under such circumstances, but we can identify whether or not they exhibit the behavioral responses expected to arise from a state of frustration.

According to Janis (1971) Freud (in 1917) was the first to postulate that aggression will always occur as a basic reaction to frustrating situations, i.e. whenever pleasure-seeking or pain-avoiding behavior is blocked. This postulate has been firmly established through clinical and experimental evidence, Janis (1971) indicates, although now it is realized that not every aggressive act is preceded by frustration nor does every frustration result in aggression. Frustration produces an *instigation* to aggression but, if the instigation is weak, it may not evoke actual aggressive behavior. But a powerful, repeated or accumulating frustration almost invariably *results* in some form of aggression, not always or necessarily expressed directly.

A widely accepted definition of frustration is 'a condition that results from the interruption of an organism's habitual sequence of acts directed towards the attainment of a goal' (Janis 1971:151). This definition involves the presence of two key components, Janis explains: an aroused need, drive or tendency to action (a motive) and a barrier preventing the motivated organism from reaching the desired goal. The feeding methods used at the Gombe and Mahale Research Centers, in which the observers restrict access to desired, elsewhere unobtainable foods for which the chimpanzees have developed a taste, contain these components.

An understanding of the relationship between frustration and aggression was developed through experimental laboratory studies on animals by a group of psychologists at Yale University in 1939. In 1941, an experiment which firmly established the essential features of a frustration situation was carried out by Barker, Dembo and Lewin. These researchers sought to demonstrate the effects of frustration upon behavior by comparing the separate behavior of individual human children involved in a nonfrustrating, free-play situation with the children's behavior in a frustrating situation. It was hypothesized that the children would exhibit regressive behavior through frustration and, as anticipated, during the experiment their behavior did, indeed, regress.

Barker *et al.* (1943:441) use the term regression to refer to 'a primitivization of behavior, a "going back" to a less mature way of behaving which the individual has outgrown,' (but not a retracing of the individual's past). Some psychologists argue that regression retraces the line of development; others, including Maier (1961:107), that regressive behavior becomes 'dedifferentiated,' but not necessarily in the same sequence or stages that behavior develops. The first view, as a reversal of normal development, permits regression to be understood as motivated, i.e. the individual returns to an earlier form of behavior which was more satisfying than the present condition. But Maier (1961) argues against an explanation of regression in terms of motivated behavior. He insists that there is a clear-cut distinction between frustrated and motivated behavior, and that the two types must be treated as *separate processes*. Maier's hypothesis and argument will be outlined, but first the parallels between the Barker *et al.* experiment and the Gombe feeding methods will be examined.

Barker et al.*'s experiment*

The experimental situation was as follows. Thirty (separate) preschool-age children accustomed to playing with certain familiar toys (set A) in a preschool laboratory situation were encouraged to play freely with several very attractive new toys (set B) for a few days. This period of nonfrustrated, satisfied free play, designated the 'prefrustration period,' was designed to develop in the child a motivation toward a highly desirable goal which she or he could later be prevented from reaching: a 'prerequisite to creating frustration,' Barker *et al.* (1943:445) explain. This was not, of course, the object of the early years of supplying the Gombe and Mahale chimpanzees with ample bait foods, freely available. But, in effect, these apes, too, experienced a prefrustration period of free access to attractive new objects – the bait foods.

It is now generally understood that a state of deprivation 'involving the lack of satisfaction of hunger, thirst, sex, or any other drive or motive – is not sufficient to constitute a frustration' (Janis 1971:155). The arbitrary nature of the imposed deprivation, and the amount to which the deprivation is effective, affect the strength of aggressive response. 'Anger is much more intensely aroused if an interference is perceived as unreasonable or arbitrary than if it is regarded as an expected or inevitable barrier,' explains Janis (1971:156). In general, situations involving unanticipated, unpleasantly surprising frustration give rise to stronger reactions than do anticipated frustrations. The experience of knowing that a desired food is present yet not available through an individual animal's own efforts – but irregularly, and surely incomprehensibly, accessible – can be assumed to be a situation involving unpleasant surprise and frustration for the provisioned chimpanzees, in that such a circumstance is never encountered in the natural food search.

After a period of free access to the new toys long enough for the child to become 'thoroughly involved' in playing with them (i.e. a motivation was aroused), a wire-mesh barrier was lowered between the child and the desirable toys. The children could still see the toys, but they were suddenly inaccessible. This Barker *et al.* (1943) refer to as 'the frustration period.' As the experimenters expected, the frustrated children reacted with anger and aggression. There was a discernible rise in tension and a change in quality of emotional expression from generally happy (positive) to unhappy (negative) expressions of emotion. Some children kicked at the barrier, threatened the experimenters, or tried to circumvent the barrier in some way. Others were less persistent and shortly retreated from the barrier.

Similarly, after periods of approximately 2 years' free access to the bait foods, a barrier was suddenly introduced between the motivated Gombe and Mahale chimpanzees and their bait-goal. The obvious barrier at Gombe is the closed, human-controlled feeding boxes. But at another level, at both centers, it is the arbitrary withholding (for a time) of access to the desired, visible or smelled foods. The subsequent aggressive behavior and the 'food-demanding displays' which Nishida (1980) reports suggest that chimpanzees are highly motivated towards a food goal and greatly frustrated by being blocked from attaining it.

According to Maier (1961:127), 'any pressure exerted that drives the organism toward the barrier hastens the transition from the state of motivation to the state of frustration.' Barker *et al.* tested the children separately; they did not seek to establish any group response to a frustrating situation. The Gombe and Mahale apes, however, were blocked from a scarce, desired food item while in groups. Having the pressure of other chimpanzees and baboons also waiting for undeterminable periods of time, for literally the same inadequate supply of bait foods, can be expected to have added the pressure of anxiety to the situation. *The presence of others waiting for the same supply of food induced direct, aggressive competition for the special bait-food among a primate species adapted to indirect competition.* The gathered chimpanzees, like most of the separate children, reacted with anger and aggression – not against the barrier, but by turning on each other (and the competing baboons) in sudden attack.

Anxiety is a normal first response of humans to danger, threat or loss (Hamilton 1982). The central feature of an anxiety state is a change of mood. Anxiety shows itself in a continuous state of tense apprehension and as irritability, 'outbursts of temper at the slightest frustration' and 'an increasing disinclination to take part in normal activities' (Hamilton 1982:4).

The most noticeable regressive behavior among Barker *et al.*'s (1941, 1943) children was a shift from complex, constructive play to less constructive or more infantile forms. Trucks formerly used to transport blocks and dolls were banged and pushed about aimlessly, dolls formerly used to play housekeeping were dangled or examined superficially. Barker *et al.* (1943) concluded that the emotional disturbance of frustration interferes with attention, planning, thinking and other constructive mental processes. It distracts the individual suffering frustration from conventional constructive pursuits. The behavior of the Gombe and Mahale chimpanzees which followed the introduction of the human-controlled feeding methods is socially primitive in that much positive behavior that supports the peaceful, successful social order changed radically and negatively, or disappeared, to be replaced by socially destructive behavior – the kind of behavior to which Maier (1961) refers as behavior without a goal.

Behavior without a goal

Psychological knowledge regarding frustration has expanded greatly since Barker *et al.*'s (1941) experiment. Today, some psychologists still view frustration behaviors as being motivated, resulting in increased vigor of goal-directed actions. Others believe that the increased vigor might be a symptom of increased aggression.

Janis (1971:148) suggests that a moderate amount of aggressive action may be adaptive, if it enables people to circumvent or eliminate whatever obstacles are producing frustration, 'as when a group of indignant employees asserts its rights to a person in authority in a way that induces him to withdraw his unwarranted demands.' Janis's view seems to define assertion, a positive, forthright statement or declaration, as being necessarily somewhat aggressive. Assertion need not contain that element at all, although it is easy to slip from positive firmness to aggression.

When the situation is such that an individual's anger mounts to a very high level, his or her actions tend to become somewhat disorganized, diffused and impulsive, reflecting an indiscriminate quality. When this happens, any resulting aggressive action is 'no longer likely to be adaptive,' Janis (1971) continues.

In making a clear distinction between motivated and frustrated behavior sets, Maier (1961) suggests that the first set is based on reason, the second on feelings. He disagrees with the argument that frustration is motivated and goal-directed. He points out that this view is inconsistent with experimental facts and that frustration behavior is nonconstructive and without a goal. Maier does not suggest that the frustrated human (or animal) lacks a goal, but that it is the 'inability to reach the goal that induces frustration and causes him to behave in a manner that leads him farther from it' (Maier 1961:Preface). He argues that the logical process is a tool of motivation, unavailable to the individual at times – or in areas – of excessive frustration. In such circumstances, when 'normal motivating processes are replaced by a condition of frustration, behavior must undergo a distinct change' (Maier 1961:128). Often, the frustration-instigated behaviors are 'just the opposite' of normal motivation-induced behavior (Maier 1961:129). 'Normal, constructive problem-solving behavior that is characterized by choices influenced by goals . . . will be replaced by nonconstructive activity such as fixation, aggression, regression and possibly resignation' (Maier 1961:128).

De Waal and Hoekstra's (1980) findings, resulting from an investigation of the contexts and predictability of seemingly spontaneous aggression among the male chimpanzees in a colony in Arnhem zoo, appear to support Maier's hypothesis. Sexual competition was not a main proximate cause for male aggression, nor was the competition for access to edible and nonedible objects such as leaves, acorns, tires, a seat out of the rain, and so on. Often the apes seemed frustrated and became aggressive when invitations to positive social contacts, including play, grooming and copulation, were refused.

A great many male conflicts seemed to arise spontaneously, without any reason that the closely monitoring scientists could detect. Males were seen to seek out their prospective opponent and attack 'without any apparent relation to the latter's previous behavior' (de Waal and Hoekstra 1980:932). The attacked ape might have been sitting quietly out of sight of the attacker, not interfering with the attacker in any way that the researchers could interpret.

De Waal and Hoekstra found that they could predict aggressive behavior appearing, following interactions that they judged intuitively to be 'very unpleasant or frustrating' for at least one of the interacting chimpanzees. In most of the 40 instances of predicted aggressive behavior which followed a frustrating incident, the action taken by the aggressor seemed ineffective to the scientists as it did not serve to 'induce the invited partner to comply more willingly' or to give up an object (de Waal and Hoekstra 1980:935). Such behavior does seem, as Maier (1961) suggests, just the reverse of a normal, constructive way of encouraging positive social interactions. 'The usual effect of an aggressive action is a dissociation between [the involved] individuals' (de Waal and van Roosmalen 1979:55). These captive apes show a strong tendency to reconcile, however; in the wild state overmuch aggression might result in permanent dissociation. This possibility will be discussed in Part 7.

Maier uses the terms resignation and apathy as synonyms. He argues that, since the trait of resignation is one of giving up of all needs and plans and a loss of hope, it is clearly a class of behavior that is not oriented toward goals. Maier (1961:34) defines the term fixation as designating a strong and persisting abnormal response, more persistent than those formed through the usual methods and learning, i.e. contiguity and reward. The fixation behaviors produced in experimental animals are characterized by (a) repetition without variation, (b) a resistance to change, and (c) a compulsiveness that is not found in a learned response (Maier 1961:77). For these reasons, Maier uses the adjective 'abnormal' to describe the rigid (rather than stable) responses. Fixated behavior is found 'to occur primarily under conditions in which the situation was made frustrating' (Maier 1961). In this type of behavior, there is a lack of ability to abandon a mode of behavior, whether or not that behavior is preferred by, or rewarding to, the fixated individual.

The possibility that the recent behaviors of the Gombe (and Mahale) chimpanzees are frustration-induced responses requires much further investigation before we can come to any firm conclusions. Nevertheless, it seems to offer a useful new perspective, worthy of further investigation. For example, why the Gombe and Mahale apes continue to wait, sometimes for hours, to compete for an insufficient ration of a desirable bait food, when the forests around them are full of a variety of wild foods, easily obtained, is a baffling question. However, when we consider their condition to be one of frustration and their behavior as fixated, their continuing to do so can perhaps be better understood (see also Part 7).

Unfortunately, because artificial feeding does not create any behavior that is completely new to chimpanzees, it is widely assumed that it changes only the *quantity* of some already existing patterns, not the quality or normalcy of behavior (Nishida 1979; Bygott 1979).

It is true that the responses induced by frustration are not new behaviors, but they are elements of a set of behaviors that is 'normally dormant' (Maier 1961:129). When circumstances of excessive frustration arouse the frustration process, it 'dominates the expression of behavior' and normal, constructive behavior is eclipsed (Maier 1961). Relief of frustration – an unadaptive attempt at a solution, involving unadaptive modes of behavior – becomes the goal, rather than the satisfaction of a need or desire, Maier argues.

Typically, there is a fluctuation between the frustration and normal motivation processes which Maier theorizes. Alternating between attacking without reason and protecting or comforting the victim demonstrates such fluctuation. Such equivocality of behavior would be expected to cause emotional distress to both of the actors, were they human. As chimpanzees share a very similar psychology, it seems reasonable to assume that it causes a similar emotive response in the apes. Aggressive action must be regarded as maladaptive if the action results in more rather than less suffering for the aggressor or for others with whom he or she is affiliated.

Certainly, the reports from Gombe and Mahale studies of aggressive, dominance-seeking, territorial chimpanzees are opposite in quality from the naturalistic studies, all of which record nonaggressive, nonterritorial, nonhierarchical animals. As Maier (1961:12) indicates, a difference in quantity or degree 'implies that the same laws and principles operate with consistency.' When a different set of laws and principles begins to operate, the difference is qualitative.

What the totally unnatural methods of artificially feeding the Gombe and Mahale apes did was to introduce anxiety, tension and frustration over food, which pushed these animals out of their optimal solution to the problem of resources competition; from indirect competition (which has none of the above negative connotations) into more costly direct competition.

Competition theory

Although direct competition is an approved social norm in industrial societies, it is a highly disapproved of and negatively sanctioned behavior among human foragers who live by the immediate-return system. Direct competition mitigates against the equality and lack of possessiveness that characterize these egalitarian societies, and it threatens the necessary peace of the group.

Competition theorists hold that being a part of a directly competing group heightens latent hostility toward others and promotes self-serving behavior (Mackal 1979). Indeed, as a result of a series of experimental studies among 10 year old Anglo-American children, Nelson and Kagan (1972) find that not only can 'the competitive spirit' be quickly roused, but it can become so overwhelming (in adults, as well as in children) that it becomes irrational. For instance, people may continue to compete, knowing they are sacrificing a reward that they could gain by cooperating, provided their action will assure that by doing so their opponents also lose the reward. Nelson and Kagan found that their colleagues had no trouble in identifying many adults in fields such as athletics, politics, business and academia, whose drive to compete over-rode their self-interest. The tendency for people to continue to compete in conflict-of-interest situations, in which mutual assistance would have rewarded both parties, often interferes with the competitors' judgment and capacity for 'adaptive cooperative problem-solving,' Nelson and Kagan (1972:56) submit. These scientists found such irrational competition among urban children in Canada, Holland, Israel, Korea and the USA. In contrast, rural children in all cultures studied were more cooperative and less willing to compete against each other. 'Ten year old children in rural Mexico cooperated, and got the prizes that eluded their American counterparts' (Nelson and Kagan 1972:53).

There is much irrationality in the recent direct competition among the Gombe and Mahale males. Their aggressive competition to gain dominance or alpha rank is irrational, in that there is 'no selective advantage in being dominant per se' (Bygott 1979:425). The competition is stressful, costly in terms of energy, and dominance rank does not correlate with copulation frequencies. The aggressive attempts of the males to deny other male group members sexual access to estrous females are irrational in that this too is stressful, and costly to all involved. Also usually futile, according to McGinnis (1979), in that, while the possessive male is engaged in driving off one competitor, other males grasp the opportunity to mate with 'his' female. Recent attempts of the Gombe and Mahale males to patrol and maintain a closed home range are irrational in that a more efficient use of the type of habitat (patchy distribution of food) in which they live would be maintenance of open groups and foraging ranges.

Mackal (1979) suggests that competition theorists find that individual competitive goals tend to be more material than cooperative goals, hence they do not serve as group cohesive measures. Since common group effort cannot realize such individual goal sets, group (social) goals are neglected, Mackal further suggests. Relations between individuals competing directly for the same objects tend to take on a negative tone and may become openly unfriendly or hostile. Since more time is devoted to self, others are assisted less. As Nelson and Kagan found, restraint of others becomes more important than self-restraint. Obstructiveness will further interfere with cooperative goals, and the group will polarize. Lack of friendliness and of coordination of behavior in friendly subgroups will maximize differences and divisions between members, and so reduce the likelihood of obtaining solutions to problems, they point out (Nelson and Kagan 1972:67).

There is considerable evidence that suggests that the Gombe – and perhaps the Mahale – groups are moving towards breakdown as functioning societies. In recent years, there has been a high level of aggressive intermale competition within both groups for dominance and exclusive access to estrous females and bait foods (Bygott 1979; McGinnis 1979; Wrangham 1979; Nishida 1983*b*; Nishida and Hiraiwa-Hasegawa 1986).

For example, in writing on the generally cooperative relationship of male chimpanzees, Nishida and Hiraiwa-Hasegawa (1986:175) cite as examples 'the high degree of tolerance' between males (seen in grooming, greeting, meat sharing, joint charging displays, food calling and opportunistic mating). But they report that there is also intense, often aggressive competition between males of the same group, to raise their dominance rank and for priority of access to 'estrous females, food and other resources' (Nishida and Hiraiwa-Hasegawa 1986:176). The tensions generated 'serve both to favor and to disrupt cooperative alliances, with complex results,' Nishida and Hiraiwa-Hasegawa (1986:175) submit. Like Wrangham (1979), they reason that, despite the rivalry and resulting ambivalence toward their peers, strong male–male bonding 'benefits each male by providing allies in defense of the community range' (Nishida and Hiraiwa-Hasegawa 1986:176). Why, then, do males of the same group also fight over dominance rank and access to resources?, the scientists ask rhetorically, and answer: 'alpha males have priority of access: achieving alpha rank therefore seems the best strategy for maximizing fitness.' 'This means that both allying with an alpha male and defeating him are adaptive strategies', Nishida and Hiraiwa-Hasegawa continue. They then comment on Bygott's (1979) suggestion that other males might not cooperate with an overly selfish dominant male, in defence of their home range. Nishida and Hiraiwa-Hasegawa (1986:176) suggest that

> an alternative argument can be made . . . . Since it is in the interests of both the alpha male and the subordinates to defend communally, the optimal strategy of the subordinates might be to cooperate even with a selfish alpha so long as he remains powerful, while waiting to usurp the alpha rank when the chance arises. Alpha's strategy, in turn, would be to form coalitions with old males [by which is meant of gamma, or slightly lower, rank] and to be wary of the beta male, who is the most likely to defeat him.

Such a strategy would surely lead to further mutual mistrust, and the polarization within the group which, Mackal (1979) suggests, may be expected as a result of direct competition. In both research areas, there is some polarization or division between adult male and adolescent chimpanzees, and almost complete separation of local groups. In addition, at Gombe, but not at Mahale, adult females have separated from the males and from each other.

In 7 years of study of the Mahale apes prior to 1977, only one direct intergroup quarrel was observed (Nishida 1979). On this basis, Nishida (1979:79) suggests that perhaps 'the rarity of the phenomenon implies the deep antagonism between unit-groups and the effective mechanism working to avoid overt clashes.' However (as is known at Gombe and strongly suspected at Mahale), the provisioned apes have eliminated a neighboring group whose home range overlaps with theirs, by killing all of the males (Mahale) or all of the members of the group (Gombe). Even if Nishida's hypothesis was correct – i.e. of a deep antagonism between groups, effectively controlled – this control mechanism has broken down at Gombe and Mahale. Furthermore, as we know, foraging theory is that open groups and cooperative use of home ranges result in far more effective use of foraging territories in which food is scattered and bunchy; hence Nishida's theory, of a pattern of antagonism between local groups acting to keep them separated, is unlikely. As mentioned in Part 2, it is Lee's (1979) opinion that the dynamic movement within and between (human) foraging groups leads to recruitment, rather than to exclusion.

Local groups of human foragers are, Turnbull (1972) points out, units in a network of a larger interconnected, interacting, interbreeding population. So, it was suggested, are local groups of chimpanzees, under natural conditions.

No naturalistic observer, including Goodall before she began provisioning, found any evidence that 'stranger' chimpanzees are attacked or chased by others of their species. To the contrary, Reynolds (1980:70) suggests that 'a potent attraction existed between the groups', while Goodall (1968*a*) and Sugiyama (1968) report that the border between wild chimpanzee groups is blurred by friendly mixing. Even the wild groups that Izawa (1970) and Kano (1971) report never mingled, showed no signs of aggression toward each other.

The Gombe chimpanzees, like the wild ones, still seem strongly attracted to encounter neighboring groups, although, in recent years, not in the positive carnival spirit of the wild groups. Goodall *et al.* (1979:49) suggest that 'the response to the sight or sound of supposed strangers, particularly the sometimes frenzied race towards the ... calls, suggest that encounters with neighbors may be extremely attractive' to the Gombe apes. 'Parties of up to ten adult males, sometimes accompanied by females and young, may patrol peripheral areas ... [of their home range], when they apparently search actively for signs of neighbors' (Goodall *et al.* 1979:48). In post-1965 years, when the Gombe apes met a fairly large group of 'strangers,' both parties tended to engage in hostile visual and vocal displays which usually ended either in mutual retreat, or through one party, usually the larger, charging and chasing the other away. When lone individuals or very small groups of outsiders are met or located through their calling, they are often attacked by the group, 'particularly if they are females' Goodall *et al.* report. (This is not male–male warfare; see Goodall 1990:101). 'Often, after a long silent patrol,' on returning to their home ground, the apes involved 'explode' into loud calling and displays, which 'gives some idea of the suppressed excitement and tension to which they have probably been subjected' (Goodall *et al.* 1979:49).

The motivation and function of this patrolling behavior at Gombe are not understood, but Goodall *et al.* (1979:50) suggest that perhaps these male excursions to the overlapping areas 'may sometimes serve to recruit females into their social group,' although this makes the attacks on encountered wild females puzzling, Goodall agrees, particularly as they do not subsequently mate with the aggressor males or join their group (Goodall 1986b:523; Kawanaka 1981). 'If then, the assaults are not directed at infanticide and subsequent recruitment of the female victims, why *are* they perpetuated?' Goodall (1986b:523) asks. The encountered females are not really 'strangers,' as we know that they had probably encountered or visited the Gombe (or Mahale) group before. But they were 'for the most part encountered in overlap zones where the chimpanzees, fearful of neighboring males, are often nervous.' 'Chimpanzees,' Goodall (1986*b*:523) suggests, 'have an inherent aversion to strangers.'[19]

The aggressive Gombe and Mahale males seem both hostile toward and fearful of outsiders but, according to foraging theory, this is not the usual or optimal response of chimpanzees in general. One of the few observed encounters of a local *unhabituated* wild group (dubbed the 'Kalande' community by observers) and Gombe females occurred in 1982. Two Gombe females had been calling for 20 minutes and, surprisingly (since they were in the camp area), four Kalande males appeared. (No Gombe males were present or had called). One of the females and her offspring quickly fled, the other rushed up a tree, chased by one of the males, who 'mildly' attacked her (Goodall 1986*b*:516). Her 4 year old son was merely 'sniffed' by another of the males, then they left. Goodall wonders if it was perhaps the strangeness of the camp setting that saved the treed female from a more serious attack. On the other hand, it takes no longer to hit or bite than to sniff, so it seems that these wild males do not share the 'inherent' aversion to strangers that is currently accepted as normal, although they too were in a tension-producing situation.

The aggressive behavior of the Gombe and Mahale groups toward encountered outsiders is a reversal of the optimal foraging and reproductive strategies, which are facilitated with maximum efficiency through the carnival meeting of wild groups. However, frustration-induced behaviors are often the opposite of constructive, adaptive behaviors (Maier 1961). While it cannot be proven, it seems possible that the warlike Gombe (and Mahale) patrols are a distortion of the seeking out of neighbors in the overlap zone for the excitement of carnival. (Either 'battle' or carnival is excitement rousing).

The phenomena of the Gombe males remaining permanently in their natal group, and of their aggressive territoriality, are the rationale for the current popularity of Hamilton's (1964, 1972) kinship, and various sociobiological, models toward understanding the social organization of chimpanzees generally. The idea that chimpanzees increase both their individual and inclusive fitness through groups of closely related males acting together to maintain exclusive access to their territory, its food resources and the breeding females of the group, is currently favored. Certainly, the mobile groups are numerically male groups, no doubt some members are closely related, and this type of group does increase the participants' individual and inclusive fitness. But the emphasis given to consanguinity is misplaced, and the aggressive and exclusive territoriality and male bonding hypotheses not applicable to chimpanzees generally. These points are enlarged on in Part 6.

Indeed, the idea of close bonds between groups of cooperating males now seems doubtful at Mahale, and untenable at Gombe, in the face of recent reports from both centers. In the early years of Goodall's study at Gombe, the typical associational pattern was the one reported from everywhere that nonprovisioned chimpanzees have been studied: i.e. the apes moved about in ever-changing, mixed-sex groups of two to six, or perhaps ten, individuals. Recent data indicate that the average size of Gombe subgroups has shrunk drastically. Working with the same Gombe population in 1975, Wrangham found mean group sizes of 1.1 in one season, 2.0 in another (Marler 1976:259). Very recently at Gombe, ambivalence between the competing males seems to have given way to alienation (estrangement) as, by 1984, 'no adult male trusts any other enough to ally' (J. Goodall, 1984, personal communication). I have not seen this development commented on directly in the publications, other than Goodall's (1990:146) remark that after his cohort Humphrey disappeared in 1979 the alpha male Figen became very anxious, as 'he had no one whom he could really trust.'

What happened to the Gombe chimpanzee society? Reports from the naturalistic field studies and of the Gombe experience demonstrate clearly the extremes in social behavior of which the same individual chimpanzees are capable, without any real change in the physical environment that would necessitate a change in the normal way of life, and certainly none in their genetic endowment. The significant shift at Gombe is in the emotional climate and social environment. Within one decade, the members of the society changed from friendly, mutually supportive associativeness to aggressive self-interest, from fluid, adaptive, charismatic authority to rigid, maladaptive, despotic power, from being a cooperative egalitarian society of self-directed individuals, and a unit of a larger society, to being an aggregation of isolated competing, individuals.

Shafton (1976) reminds us that among higher nonhuman (and the human) primates, the following trends are indicated.

(1) Selection favors behavioral plasticity, a trait which appears to have arisen from the survival value of exploiting a spectrum of types of habitat.
(2) Equal potential for aggressiveness or peacefulness are both part of the genetic constitution of all animals; however, no genetic coding *inevitably* results in either aggression or peacefulness. Rather primates possess a set of genetic attributes that can be expressed as aggressiveness under particular sets of conditions, for example, for self-defence (Dubos 1973).
(3) The manifestation of genetic potentialities is affected by past experiences (including learning) and present circumstances.
(4) We also know that an animal's whole psychology and social structure is a secondary adaptation to its primary needs to feed and to reproduce. 'On an average, social bonds will not interfere or run counter to the basic adaptation to the environment. The psychological and the social organization are the result of the adaptive syndrome to the environment' (Dittus 1975:236).
(5) While social competition is inevitable among animals (Deag 1980), it need not take a direct, confrontational form. Chimpanzees are adapted to indirect competition for food, a separate, simultaneous seeking of the same food resources. The observers taking control of access to a desired and scarce food pushed the Gombe and Mahale apes into direct, prolonged, confrontational competition with others for the bait foods, under conditions of high tension. This situation is unlikely to come about except in the extremity of a prolonged food crisis and crowding brought about by a major geophysical disaster, due to the adapted pattern of foraging separately and silently, when food is short.

Shafton (1976:8) points out that natural selection endows an animal with the genetic means for developing a set of behavior systems which 'lets it deal adaptively with a class of situation which recurs in the life of its species' in its particular niche, in both social and nonsocial environments. 'Taking behavior over time, it appears that the animal commences and terminates action only by the separate cutting in and cutting out of a coordinated array of perceptual biases, motivational tendencies, and motor potentials' (Shafton 1976:8) The different mind set which Maier (1961) claims is activated by frustration suggests the cutting in of a normally latent behavior system. The adapted social mechanisms that dispense normal stresses no longer work, with the result that the Gombe chimpanzees have changed *from system-maintaining behavior to behavior that is system-changing*.

The recent behaviors of the provisioned Gombe and Mahale apes are obviously part of the ethogram of the species, though not adaptive under the actual ecological circumstances. This raises the questions: What are the selected factors that produced the recent negative behavior of the Gombe and Mahale chimpanzees? What is the selective value in such behavior? In what circumstances might they be adaptive? These questions cannot be answered adequately in the context of this study. They require deep and lengthy consideration. But one possible explanation is offered in Part 7.

It was mentioned earlier (but bears repeating) that biologists now believe that an animal species may be social or asocial, peaceful or aggressive, depending on circumstances. The strong contrast between the negative behavior of the stressed Gombe apes and the positive social behavior of the wild chimpanzees is in accord with the biological view. Because the analysis of chimpanzee society is often suggested as a model for the evolution of human behavior, we must keep the biological understanding in mind.

PART 5

The mutual dependence system

The principles and patterns, behaviors, interactions and relationships that structure a mutual dependence system require more explanation and substantiation than they have received to this point. Accordingly, in this part of the book, evidence from naturalistic studies of chimpanzees and from field studies of the six human groups who, when studied, lived by an immediate-return foraging system, are used to develop further the model of the basic mutual dependence system as it is among chimpanzees.

In reading the rest of this volume, three earlier mentioned points must be kept in mind. First, it is now generally agreed that chimpanzees share the same range of emotions, and similar mental processes, with humans, though they are technically greatly inferior.

Second, the mutual dependence system as outlined is a *model*, and a preliminary one at that: a first attempt to understand the adapted social structure of chimpanzees from a new perspective. In reality, chimpanzee social groups will vary from the model. Chimpanzee social behavior, like that of humans, is extremely plastic. Differing ecological conditions will affect local behavior and local traditions. For example, Sugiyama and Koman (1979) found that in the wild group that they studied for 6 months in the vicinity of Bossou, Guinea, one particular male leader, 'Bafu', made repeated attempts to keep the entire Bossou group together, going back and forth making barking calls as they traveled, apparently trying to keep together the forerunners and those lagging. This is leader behaviour quite different from that of the model.

A third point to keep in mind is that most observers studying chimpanzee behavior and organization think in terms of dominance and aggression, and lack of these qualities. But pacifism is a specific state which must be maintained, not simply a vacuum that occurs when aggression is missing. In this section, a start is made toward understanding the checks and balances that maintain the necessary peace which is integral to the adapted way of life.

The model of the mutual dependence system was outlined starkly in Part 1. Part 5 is devoted to developing the mutual dependence system as fully as possible, using human-based theories that seem to apply and within the limits of the available evidence. The reader is reminded that the premise set forth is that this adapted social system is based on an ecological adaptation, i.e. a fusion and fission foraging pattern of open groups around open home ranges, in indirect, non-confrontational competition for resources.

To reiterate, the mutual dependence system is structured around: (1) a very fluid mutualistic relationship between individuals who take shifting, mutualistic, charismatic and dependent status/roles; (2) mobile and sedentary (rather than male and female) subgroups; (3) local groups as units of a larger interacting society; (4) personal autonomy or self-direction, yet a high degree of conformity; and (5) an extremely high egalitarianism.

The nature of authority

Authority is inseparable from the roles, statuses and norms of a society, despite the informality of some forms that sometimes creates the illusion of the nonexistence of authority (Nisbet and Perrin 1977:106, 109). Certainly, organization based on the principles listed above does give an appearance of no organization, to the observing researcher. But every continuing society, animal or human, has some form or pattern of authority, even if difficult to detect. Even where the authority is implicit, and derived from ancient mandates, a member of any society obeys a number of clearly understood (socialized) ordinances, Nisbet and Perrin (1977) point out. The moment two individuals find themselves 'in a group or relationship that involves, in whatever degree of informality or formality, the distribution of responsibilities, duties, needs, expectations, privileges, and rewards, a pattern of authority is present' (Nisbet and Perrin 1977: 106).

Merely to follow the specifications of a role is to engage in and be part of a pattern of authority. Both charismatic and dependent chimpanzees are volitional bearers of rights and role obligations – implicit ancient mandates – that are adaptive toward their survival.

Ethologists use different definitions of role when referring to animals and to the human species. The essential difference seems to be based on the instinctive nature of much animal behavior and on the cognitive nature of most human behavior. Because I write in functional terms of statuses and roles in a particular type of social organization which is characteristic of foraging humans and at least one ape species, the chimpanzees, I use a very broad, simplified anthropological definition. In this study, *roles* are defined as being expected and sanctioned behaviors structured around specific rights and responsibilities which indicate a particular status within a group or social situation.[20]

For the same reasons that I do not define role differently for chimpanzees and for human foragers, I do not operationalize the concept of status. It, too, is broadly defined in the same way for both human and chimpanzee societies. As used here (adapting Kemper's (1978:34) sociological definition slightly), *status* is the reward and recognition which others give to an actor voluntarily, without coercion, in response to the actor's competence in social relationships. A *status position* is a defined position in the social structure of a group or society that is both distinguished from, and related to, other positions through its designated rights and obligations. It does not necessarily imply hierarchical position or rank, but is expressed in terms of a role. A dog trainer, a widow or a husband are examples of statuses. So is a charismatic leader, or a dependent follower, whether human or chimpanzee. These are not ranks. A rank involves a lineal arrangement or standing, relative to that of others, as in a hierarchy.

Humans are conscious of their roles in society. It is not yet established whether chimpanzees are, but de Waal supplies evidence in terms of the observed behavior of the Arnhem zoo chimpanzees which does seem to indicate that the animals have a concept of how to comport themselves within the social definition of a particular role. Take, for example, the role behaviors essential to being accepted as a charismatic leader, which is discussed fully later. Briefly, what is required of the animals taking this role is that they be assured, approachable protectors of the vulnerable and keepers of the peace within the group, and that they generalize this approachability and acceptance of others widely, rather than having fixed alliances with certain preferred individuals.

De Waal (1982) writes of a male chimpanzee, Luit, who challenged, and for a time won alpha position over, an older male, Yeroen, who was a long-established and group-supported charismatic leader of the chimpanzees in the Arnhem zoo. When he achieved alpha position, Luit – who until that time had been a 'winner-supporter' in aggressive interactions, joining in with the winning animal to attack the losing individual – suddenly and clearly 'adopted a brand new policy' (de Waal 1982:123–4).[21] 'Luit's social attitude altered totally' (de Waal 1982: 124). He became a 'loser-supporter', a protector of the vulnerable. When serious quarrels broke out, Luit now leapt in and literally beat the battling animals apart. 'He did not choose sides in the conflict' as formerly. Any animal that continued to fight received a blow from Luit, who intervened impartially. De Waal comments that he had never seen this male act so impressively before. In less serious squabbles, he was seen to put his hands between the quarreling pair, force them apart, and stand between the antagonists until they calmed down. (Human foragers are also seen to intervene in this manner, to stop fights (Lee 1979: 380)).

Other Arnhem chimpanzees intervened in aggressive incidents on the side of the antagonist with which they were most familiar and friendly. Yeroen and Luit were the two exceptions (de Waall 1982). These apes did not let personal friendship affect their peace-enforcing interventions, but intervened impartially between any disturbers of the peace. Moreover, it is de Waal's (1982) impression 'that Yeroen and Luit are extremely aware of the functional effect of what they do,' which, to de Waal, indicates rationality. He suggests that these males intervene in accordance with a policy directed towards increasing their power. I view this as the role behavior necessary to charismatic leaders, although in the wild societies such aggressive encounters rarely break out, owing to effective social controls including the tendency of individuals to leave any situation which is becoming tense. Yeroen and Luit's fulfilling this essential role can be expected to increase their influence with, rather than their power over, the other members of the group.

Bernstein (1964, 1966) conducted experiments in connection with the 'control' role among caged groups of several species of monkeys. Several times he removed the control animal, which resulted in another group member immediately taking on the 'control' or leader behavior, although this animal had never before exhibited this behavior. This suggests that a control role is an adaptive aspect of primitive behavior. But does not the suddenness of the change of policy on Luit's part, when he achieved dominance over the elder male, suggest also a consciousness of the role behavior expected of a charismatic leader? It seems that Luit had an understanding of the functional role expected of charismatic leaders and enacted it as conscious behavior.

One might not expect the authority structure in a highly egalitarian society to take the form of leaders and followers, but the combination is made possible through the aforementioned patterns of fission and fusion, indirect competition and autonomy of adult individuals – in short, through the freedom to choose. Not only do chimpanzees choose whether to stay with a group, or to leave, but they choose also whether to be part of a charistmatic-dependent alliance or not: with whom, when, for what period of time. By choosing their companion they choose the role they will assume for a time – charismatic leader or dependent follower. The common experience of adults shifting easily from one role to the other maintains egalitarianism and mitigates against the formation of ranks, classes or elites.

The ultimate bases of the authority structure are two: the parent–child role relationship which, generalized between adults, becomes the charismatic–dependent relationship; and the inborn tendency of primates toward a variability of temperament, which in turn inclines the bearers toward charismatic parent or dependent child role, in the generalized adult relationship. Before we consider these principles, the natural parent–child role relationship is discussed, because the parental role and contribution of the male chimpanzees is much underestimated.

The parental roles

Although it is not the way modern biologists measure fitness, the successful raising of healthy offspring to the age of reproduction is what Darwin meant by the term 'reproductive success.' Reproductive success, as Darwin defines it, cannot be measured in terms of the male contribution only. Only among species whose reproductive responsibility extends only to the producing of a mass of fertilized eggs (turtles, for example) can individual reproductive success be measured in terms of the number of females with which a male manages to copulate exclusively.

It is generally agreed that the parental investment of the female chimpanzee in time and energy spent to raise an offspring to reproductive maturity greatly exceeds that of the male parent. Ghiglieri

(1984:179) suggests that the parental investment of the male chimpanzees may amount to only 'the time and energy spent in courtship and insemination.' To the contrary, in terms of raising healthy offspring – the offspring of the group, not solely their own biological children – the parental investment of the male chimpanzees is important and perhaps in the long run as energy-consuming, if not as personal or intense, as that of the mothers. For example, I view the seeking of food that is ripe and ready to consume and the food-calling behavior of the mobile males on finding it as essentially parental behavior, both natural and generalized. It is their contribution to the feeding of the young whose mothers respond to the summons, and of the dependent and vulnerable members generally, i.e. the old or the ill for whom the food search is more costly in terms of their lower energy level.

Hrdy (1981:76) writes of the 'situation-dependent' character of relationships between primate males and infants. She suggests that, when the mother is around to take care of the infant, males may seem aloof, but that this appearance of disinterest belies the males' critical role in emergencies. Certainly in the normally unthreatening natural habitat of wild chimpanzees, the protective aspect of the male parental role is seldom operative. Little direct, overt action is called for in this generalized fathering role; hence the observable behavior of the adult males appears to be one of a tolerant lack of responsibility toward the young, each of which is intimately nurtured and protected by its own biological mother. Nevertheless, despite the lack of clear, observable demonstration of the role responsibilities of the 'fathers', the males normally have a strong, broad, generalized responsibility toward, and interest in, the welfare of the young.

Russell and Russell's (1972) hypothesis is that under relaxed natural conditions in which space and food are ample, the parental behavior of non-human primates takes two forms. There is a *social* form of positive parental behavior generalized toward the whole group, which is usually the male role, and an intimate one-to-one nurturing and caretaking role to her own biological offspring, which is the parental role of the mother. These are interdependent parental roles and the evolutionary stable reproductive strategies of the male and female chimpanzees. The male takes on the generalized parental role, i.e. the welfare of all the young, among them usually his own, although he is probably not able to identify them.

The pattern of one biological mother/many social fathers is a more adaptive arrangement than that of two biological parents, for the survival of young chimpanzees. The males do not recognize genetic relationships, nor are they responding to an abstract concept of family. They respond as protectors of all the young and, even more broadly, of all the vulnerable members of the group. Not having knowledge of, and foremost interest in, their own biological children frees male apes to take a protective responsibility toward all their (social) children. Probably the unconscious, ultimate motivation is that the only way that guarantees that each individual male's biological offspring are protected is for the male parental role to be generalized. Only by his protecting all is it certain a male's own offspring are protected. Males must, of necessity, be 'protectors-general.'

The females are especially charged with the welfare of their own biological offspring, so a very close bond between individual mother and child develops, on a one-to-one basis. When mothering, the female is a 'protector-specific.' It is probably because of the strong bonding to one particular female, their mother, and the much more generalized bond with all of their 'social fathers' that the young appear to be more strongly attracted to males in general than to other adult females, a tendency on which many researchers have commented.

In the literature one reads of the difficulty of locating the 'timid mothers.' However, as Kortlandt (1962:129–30) points out, caution is the most conspicious feature of maternal behavior, and 'the mother's main task seemed to be to avoid exposing her offspring to any risk whatever.' The necessary role behavior of the mothering chimpanzees, in their intimate concern with the welfare of their own personally dependent child, is to retreat. This is their adapted mode of child protection, not a sex-based or personality trait. Female chimpanzees have the same wide variation in temperamental and personality traits as do the male apes.

Sugiyama (1972:158) mentions that, when he encountered a group of the wild apes, 'one or more of the big males displayed against him by exposing their whole bodies, vocalizing, and brachiating from branch to branch,' while the rest of the group hid silently in the foliage. The 'rest of the group' included mothers of young, and timid animals of both sexes, rather than, as is usually assumed, the female apes, with and without young. Nevertheless, both displaying 'fathers' and hiding (cautious) mothers are fulfilling their particular protective parental roles. In time of danger it is the role-task of the protectors-general to act in a way which will draw attention to themselves and away from the young which are in the care of their mothers, or to protect more directly the vulnerable mothers and young. Hence, it is mainly males that display and threaten attack. The mothering females are not timid by sex or nature, but *retreative within the requirements of their vital parental role.*

It is obvious that, if a mothering female should take the more aggressive protective role and be seriously wounded or killed, her infant's chances of surviving are diminished. The infant's probable mortality is little affected by the loss of her or his biological father. Just distracting attackers without removing the young from the vicinity, or vice versa, is less effective than both sexes cooperating in their particular parental roles. This mutually supportive parental relationship which affords maximum protection for the young, not numbers of copulations as such, will lead to (Darwinian) reproductive success.

The parental strategies are ultimately sex-based, but the generalized 'social' father role may be taken by any adult or near-adult not engaged in the mothering role. Human foragers live in nuclear families, but, although both parents know and care for their own biological children, there is also this underlying primate parental pattern. Among human foragers, adult males are both biological and social fathers, but all social fathers are not male. It may be recalled that the Mbuti foragers' infants, before they can walk, are allowed to crawl all over the camp. Turnbull (1978) assures us that if the roaming infant gets into difficulty, any adult will take care of it. It may be an unrelated young or old man or woman, who 'mothers' the infant. Turnbull (1978:177) suggests that this generalized care builds the child's sense of security, for he or she learns early that there is a 'plurality of mothers and safe territories.'

Goodall (1971*a*) supplies an excellent example of similar generalized parental behavior among the Gombe apes, which is outlined in *ibid.*, p. 153. She writes of watching the gentle caretaking actions of a group of adult animals resting in the forest where, she suggests, they are more relaxed than in the camp area, toward a very young 'still tottering' infant (Goblin). The broadening of the attachment of the chimpanzee's child from mother to group also begins early (Goodall 1968*b*:372).

Under normal conditions, and even under stress, male chimpanzees are often affectionate to infants, patting or hugging them, and amiably tolerating the infants' pulling and climbing on them. Like many human parents, chimpanzees tend to refrain from asserting much overt authority towards their child as long as the child is conforming, not endangering itself or others. Yet authority is implicit in the relationship and can be quickly asserted if the circumstances or the child's behavior changes in an undesirable direction. From the published evidence it appears that the chimpanzee social 'fathers', as well as (sometimes more than) the biological mothers, are disciplinarians. Goodall (1975) writes of a Gombe infant 'pestering' a baboon by kicking at him, whereupon the baboon bit him and the infant screamed. A young male chimpanzee charged and hit the baboon, but minutes later the infant was again kicking the baboon. This time the young male gave the infant a hard slap, and the child, subdued, went to his mother.

Goodall recognizes the disciplinary nature of such intervention. She submits that 'adult males are, in general, protective of all infants (and may discipline them, too)' (Goodall 1983*a*:53). In keeping with the general familial nature of egalitarian relations (as seen in the human foraging societies), all older members of the social group show almost unlimited tolerance toward, and are protective of, any infant who has reached an age to venture away from its mother (and so has begun to enter the larger community). Thus the behaviors expected by the group are impressed on the maturing child.

In all societies, adults have higher status and rank and more authority than children. Accordingly, a varying amount of deference is expected from children as they mature socially. An age-based deference appears to be expected and enforced (taught) by wild adult chimpanzees, although such recorded incidents are usually viewed as being dominance interactions by the observers concerned. Goodall (1967*b*:168) writes of Flo's daughter Fifi's adjustment to juvenile status, 'learning what liberties she could take with the leaders and which would bring a sharp reprimand.'

Albrecht and Dunnett (1971:35) report that 'large chimpanzees of any sex and class tended to dominate smaller ones.' To back up this statement they cite an incident in which a tree-climbing mature male hit out at an adolescent, which was also climbing, making him give way. In the wild group that he studied, Sugiyama (1972:151) observed young chimpanzees moving off a branch or trail, out of the way of an approaching older animal. Following the then-established paradigm, he, too, interprets this behavior as 'mild dominance' or 'aggressive–submissive' interaction.

Sugiyama (1969:199) describes as the mildest type of aggressive dominance 'a confidence gesture,' in which a larger (older?) 'male' walks with 'exaggerated composure' up or near to a smaller (younger?) individual whereupon the latter animal crouches, 'sways back grinning' or moves away. The second level of aggressive dominance, 'a light attack,' Sugiyama suggests, consists of a large male staring at a small male or juvenile who has 'inadvertently approached him heedlessly' (Sugiuama 1969:199). Sometimes the larger ape threatens the younger by bobbing or slapping a branch, whereupon the startled offender usually fear-grins, screams or runs away.

The third form of behavior considered by Sugiyama as being aggressive in tone or nature is chasing. Again, this interaction is usually a matter of larger animals pursuing smaller ones; however, the scientist notes that the larger ape's action looks exaggerated, as though it were 'a false attack' (Sugiyama 1969). MacKinnon (1978:79) points out that chimpanzees have 'very stereotyped displays and controlled attack behavior which enables them to settle disputes with a minimum of injury,' thus reducing real aggression to a minimum.

No naturalistic observer saw a violent attack on a young ape (or on a peer). Sugiyama [1969:199) notes that, even when a large ape seized a smaller one, it did not bite or harm it. In fact, with two exceptions, the 31 incidents observed over 6 months which Sugiyama (1969:200) labels as 'dominance and aggressive-submissive' behaviors seem (judged on a basis of larger and smaller size) all to be between adult and subadult chimpanzees.

Ghiglieri (1979) reports that he rarely saw examples of dominance interactions between the Kibale apes. Dominance interactions, 'if such they are,' are 'uncontested approach–displacement sequences in which an adult usually asserted dominance over an immature individual' without physical contact (Ghiglieri 1979:238). Perhaps some incidents reported as dominance may be examples of 'social parenting' – the disciplining of a juvenile ape by an adult (usually the nearest) other than its mother, teaching deference to adults, for example.

Other episodes in which the Reynoldses were able to discern the nature of the interaction include four incidents in which an adult animal chased a juvenile, plus three others between two juvenile or adolescent animals. These latter quarrels seemed to set off barking by an adult and Reynolds and Reynolds (1965:416) observed that when an adult male moved toward the quarreling young the squabble stopped immediately. Albrecht and Dunnett (1971:45) report the same phenomenon: when a playing group of young apes was acting 'rather rough', an adult female 'merely screamed' at them, whereupon they quieted down. Interestingly, Turnbull (1978:186) makes virtually the same report regarding an incident that he observed among the Mbuti pygmies. When a child playgroup became overboisterous, an adult shouted a reproach, and the children responded by quieting.

Once again, rather than being examples of aggressive or dominance behavior, such interactions may be examples of the generalized parental authority of all adults over the young; socialization – through enforcement – of the maturing young to the behavioral rules, the social norms.

Any social scientist would recognize the intervention of the Mbuti adult between the squabbling young as a conscious lesson in socialization towards a norm of peace and nonagression. Very broadly defined, norms are approved and disapproved behaviors.

Anthropologists consider norms to be a vital core of culture, more so than the use of artifacts, rituals or roles. Some primatologists feel uncomfortable with the concept of roles and ritualized or expected behaviors among primates, but recent research shows the chimpanzee to be much more a cultural animal than we had realized. It seems that if we accept that chimpanzees carry out roles and ritualized behaviors we must accept that there are approved and disapproved behaviors, e.g. norms. It is interesting, and may be significant, that Sugiyama (1969) describes the actions of a larger ape chasing a small one, and Ghiglieri (1988) the highly nervous response of a familiar female, Zira, to his slightest movement as 'exaggerated.' If the judgment of these scientists is correct, it strongly suggests that these chimpanzees quite consciously communicate messages regarding expected social behavior.

Variability of temperament

Perhaps the ultimate basis underlying the social system that has evolved among chimpanzees is an inborn tendency known to be typical of primates, a pronounced variability of temperament. Temperament, 'the raw stuff of individuality' is defined by Hall (1941:909) as 'consisting of the emotional nature, the basic needs structure, and the activity level of an organism.' Temperament underlies (and affects) character, which is in part formed by an individual's experiences, especially those in early life. Character may be changed, to some extent, through new experiences. Nervousness, aggressiveness, assertiveness, spontaneity, variability, speed of reaction and activity are examples of temperamental traits. Such traits do not change, but the character formed through experience may affect their expression.

Nissen, who in 1930 was one of the first scientists to attempt a short study of wild chimpanzees in their natural habitat, also studied captive chimpanzees at the Yerkes Laboratory of Primate Biology. As a result of both experiences, he realized that thresholds of excitability in chimpanzees vary tremendously. Some chimpanzees are characteristically 'jittery; they jump at every sound and movement. Others are calm and poised' (Nissen 1956:410). These differences are often apparent when infant chimpanzees born and raised at the Yerkes Laboratory are only 6 months old, Nissen (1956) indicates. He points out that, among the nursery-reared infants, these are differences in temperament, not related to any consistent differences in experience, because since 1939 every effort has been made to keep laboratory environment conditions uniform. Hence, Nissen became convinced that the basis of this pronounced individuality among chimpanzees is genetic.

Many field observers report marked individual differences in the behavior and personalities of chimpanzees, even of wild ones with whom they were not long familiar. They found that some apes are calm, confident and assured, others shy, timid and nervous in the same situation. For example, Goodall (1968*a*) and Sugiyama (1972) report that before the chimpanzees in their study groups became accustomed to their presence some apes hid in the trees on catching sight of the humans; others did not. Goodall (1968*a*) notes that those individuals who hid from the observers tended also to hide when violent social action took place around them.

Reynolds and Reynolds (1965:416) report that certain individuals among the wild apes showed curiosity, rather than fear, when they saw the researchers. There was a strong contrast between the suspicious, shy manner of the majority of the wild apes that Reynolds and Reynolds studied and the relaxed, confident, even fearless, manner of some others of the group. Goodall (1971*a*) and Ghiglieri (1979, 1988) report the same of some preferred companions. Ghiglieri (1988:68) writes of an 'ancient' female, Gray, whose 'calm but inexplicable acceptance' of the observer 'contrasted with the exaggerated nervous reactions' of her most constant companion, a young adult female, Zira. It may be this pair, or another, to which he refers when he reported in an earlier publication (Ghiglieri 1979) that one Kibale ape took only 1 month to become habituated to his presence, in sharp contrast to her (or his) preferred companion, who took 15 months before becoming relaxed when Ghiglieri was present.

When Goodall (1971*a*) first began her studies at Gombe in 1960, it did not take her long to recognize not only that the chimpanzees indeed had distinctive and decidedly different personalities, but also that chimpanzees with very different characters tended to spend much time together. Her popular book, *In the Shadow of Man* (Goodall 1971*a*), is an excellent source of descriptions of the contrasting characters of some preferred partners.

As we know, the first ape to enter Goodall's camp (David Greybeard) came to pick the wild fruit from a tree near which Goodall had placed her tent. The ape David was 'exceptionally calm' and gentle, with an air of natural dignity (Goodall 1963*b*:296). He was frequently seen 'calming and reassuring the other apes with a touch.' In this way he functioned as a kind of 'counselor,' Goodall (1963*b*, 1971*a*) suggests. The pioneer observer was always pleased to see this male among a group because, with David present she had a better chance of approaching, to observe the apes more closely. His lack of fear seemed to reassure more timid members of the group. Ghiglieri (1988) makes virtually the same observation regarding the calm female Gray, and Albrecht and Dunnett (1971) of the behavior and influence of a wild adult female, Pandora, whose confidence and lack of fear also reassured the more wary members of the group.

In a recent book, Goodall (1986*b*:209) comments that David Greybeard 'possessed all the characteristics that seem to be most significant for leadership.' Between 1961 and 1965 'about 75 percent of the newcomers (to the feeding station) arrived for the first time with David' (Goodall 1986*b*: 209).

The ape William was apparently the most timid male in the Gombe group. Yet this exceedingly timid animal was a frequent companion of the very confident David Greybeard. Goliath also was frequently with David. Goliath was not especially large, but of superb physique (Goodall 1971*a*:45). He was also 'wild, impetuous,' of 'uncertain temper' and easily roused to violence (Goodall 1963*b*:297). Goodall vouches that it was a long time before this ape behaved calmly in her presence. If she made a sudden movement, Goliath started, and threatened her vigorously. As already mentioned, Goodall had a large cage specially built, into which she could retire for safety when Goliath's temper was at its most violent.

Also, in the early years, Goodall (1965*a*:454) considered the ape David to be 'the dominant male of the three except on rare occasions.' She reports that if some other ape approached this gentle animal entreating attention, David was always 'quick to respond with a reassuring gesture' (Goodall 1971*a*:77). Often, too, if Goliath became agitated while in the humans' camp, David would reach out and touch or stroke him, which calmed the nervous, aggressive male.

According to Goodall (1965*a*), David was always 'dominant' over William, and also over the aggressive Goliath, except when the latter became particularly excited. Then the nonaggressive David ran from him screaming. Significantly, Goodall continues: 'Goliath was always dominant over William when David was present; when David was absent *the other two appeared to rank as equals*' (my emphasis). Goodall noted that there was a greater degree of mutual attraction between Goliath and David than there was between Goliath and William. It is on the basis of the observations that William did not take the follower role when with Goliath, that I judge Goliath, like William, to be a dependent individual, lacking the characteristics of a charismatic leader, and David a true, even model, charismatic leader. However, in her 1971 book, Goodall, apparently thinking in terms of the aggression-based dominance model, indicates that, on the basis of other apes giving way to Goliath on a trail or when in pursuit of the same single banana, she began to suspect that Goliath might be the 'highest-ranking' (dominant) male chimpanzee in the group. Later she found 'that this was indeed the case' (Goodall 1971*a*:76). She comments that she had great difficulty in determining David Greybeard's position in the (forming) dominance hierarchy, because calm and gentle David was neither aggressive nor submissive. David avoided fights, but when roused, he was fearless. It was David who led 'the first determined joint challenges' against the alpha male Mike (who deposed Goliath), Goodall (1986*b*:61) submits, which resulted in the overthrow of this despot.

Two other very dissimilar Gombe apes who spent a good deal of time together in the early years, are the elderly females dubbed Flo and Olly by Goodall. Goodall (1971*a*) describes Flo, the first female to come to her camp feeding area, as being very old and frail in appearance. But 'her character by no means matched her appearance: she was aggressive, tough as nails, and easily the most dominant of all the females at the time' (Goodall 1971*a*:88).[22] The old female, Olly, who often traveled about with Flo in the early days, was 'remarkably different' from Flo. She was tense and nervous in her relationships with the other apes, particularly so with adult males. Goodall (1971*a*:89) comments that Olly's 'frenzied pant-grunts rose to near hysteria' if Goliath approached her. Quite differently, Flo was usually 'relaxed in her relations with the adult males' (Goodall 1971*a*:81). During the early years at Gombe, Flo was often seen grooming in a close group with several males out in the forest, and in camp (before 1965) she showed no hesitation in begging bananas along with the males.

Goodall (1971*a*) lists other such 'opposite personality' combinations. Leakey and Mr Worzle were two mature Gombe males who were very different in temperament, yet who 'spent hours in each other's company,' grooming, feeding or traveling together (Goodall 1971*a*: 126). Leakey was robust, 'high-ranking' and usually good-natured. Mr Worzle, a sickly animal (thought to be Leakey's brother) was always nervous, both with the chimpanzees and humans. In the aggressive (post-1965) Gombe society 'he was very low-ranking' indeed, Goodall (1971*a*) vouches.

The chimpanzees whose temperaments are described above probably captured the various observers' attention by reason of being unique, vivid personalities, extraordinarily confident or nervous, in contrast to the majority of more average individuals, whose temperaments and personalities were less outstandingly one extreme or the other. Shils (1965:200–1) reminds us of Weber's (1957) suggestion that some extraordinarily charismatic persons may impose themselves on their environment by their exceptional 'courage, decisiveness, self-confidence, fluency, insight, energy, etc.' through qualities of personality and behavior (without any implication of divine endowment). The common feature that such charismatic individuals share is their extraordinariness, Shils (1965:201) indicates, which is constituted by 'the high intensity with which certain *vital, crucible* qualities are manifested' in contrast to the low intensity with which they appear in more ordinary individuals, who possess a 'watered-down' '*normal charisma*.' The extraordinarily charismatic individual embodies, or is seen as embodying, something 'serious' in the Durkheimian sense; that is, socially centred and fundamental to the group's or society's existence (Durkheim 1953). The 'more average' (majority) of chimpanzees are probably, like the majority of humans, more knowledgeable or more confident in some situations and less so in others. Theirs is, as already suggested, a more attenuated 'normal' charisma. This more common form and consequent fluctuation according to the situation facilitates easy change from leader to follower status and role by most chimpanzees, according to circumstantial needs. There seems to be no reason why Weber's insight cannot be applied to outstanding chimpanzees such as David Greybeard and Flo, and Shil's concept of 'normal charisma' to the more ordinary or average animals.

It seems that in any chimpanzee group studied there are a few more than ordinarily charismatic animals, whose personality qualities are so outstanding that they are, much more frequently than most, spontaneously assigned the charismatic leader role. (There are a few exceedingly nervous animals, who usually prefer and are assigned the follower role).

Their very definite personalities and confident demeanour make discernible the qualities and behaviors that compose the charisma role. Reynolds and Reynolds (1965:415) dubbed a group of four very calm, quite fearless elderly males often seen together 'the royal clan' (Reynolds 1965a:195). These animals were identifiable by their forceful movements and by their 'relaxed and confident bearing and unhurried gait' which made them particularly noticeable (Reynolds and Reynolds 1965:415). The confidence of charismatic leaders corresponds to sureness, lack of hesitation (i.e. decisiveness) and self-assurance. Possession of such personality traits probably make it easy for the bearer to be normally relaxed and calm – unthreatened – hence approachable and friendly.

Nervousness, which leads to dependence, may also be expressed in posture and expression. Lack of assurance and a tendency to fear the novel are signaled through the hesitancy of some of the apes in the face of anything unusual or even faintly threatening. The model dependent chimpanzee, then, is a more hesitant, more easily alarmed, less assured, more nervous (hence sometimes more aggressive and sometimes more timid) individual than the model charismatic ape, with a demeanour and behavior which reflect these qualities. But, as we know, personality qualities alone are not instrumental in the achievement or assignment of either charismatic or dependent status. Both statuses are ultimately dependent on an individual's carrying out the behaviors attached to one or the other role on which the mutualistic relationship is based.

Since charismatic leaders (and also dependent followers) are many, most have less outstanding personalities than those individuals mentioned above. De Waal names several females who are less impressive than the highly charismatic Mama, but who (in the zoo situation) head more or less permanent subgroups (see de Waal 1982:74–81). Nevertheless, the proposed *model* of a charismatic leader in the natural, wild chimpanzee society is any of a number of animals of either sex who are, to varying degrees, confident, self-assured, normally nonaggressive, but fearless when roused, tolerant of others, approachable and responsive, with a 'presence' through posture and bearing (rather than through size and strength) and who carry out leader-role related behaviors. The majority of chimpanzees, the average members, being neither outstandingly confident nor outstandingly nervous, are less easily identifiable in either role. Moreover, the average animal changes from charismatic leader to dependent follower role much more frequently than does a David or a William, a Flo or an Olly, a Gray or a Zira.

The nature of true leadership

Leadership theory based on studies of humans is usually thought of as applicable only to human leadership. Although it has not been used to do so, of course it is entirely appropriate to use this theory to analyse the authority structure of human foraging peoples. If the social system of chimpanzees takes the same form as that of some human foragers, a theory that is found to be useful towards understanding authority among these humans should also explain much about the authority structure in chimpanzee groups.

Interaction theorist Gibb (1970*a*, *b*) maintains that true leadership in human groups is not a form of authority determined by caste, class or other factors which permit a forceful insistence on obedience from inferiors. Such a power-based form of authority is 'headmanship' or 'dominance.' 'True leadership' is based on popular selection and acceptance (Gibb 1947).

Leadership theory excludes groups such as dominance orders, in which there is a head or top person in alpha position, (usually postulated to be male) because the dominant or head man may regulate the activities of others for his own self-interests' as he chooses through a forcible assumption of power. Such persons are despots. interested more in their own goals than those of the group, Gibb (1947) suggests. There is 'a wide social gap between the group members and the head (or dominant member), who strives to maintain this social distance as an aid to his coercion of the group through fear' (Gibb 1947:213).

Leadership is not a quality of personality that some persons possess and others do not. It is a social role, the successful achievement of which 'depends upon a complex of abilities and traits' (Gibb

1970*b*:207). Leadership is an interactional function of both personality and specific social situation. A leader is a member of a group, whom others of the group select and accept, and in doing so, confer a certain status. Leadership describes the role by which the duties of this status are fulfilled. Viewed in relation to the individual, true leadership is a quality of role within a specified social system; in relation to the group, it is a quality of the structure. Because leadership is a social role, the same individual in a group may alternate between the role of leader and follower as the situation warrants, Gibb (1947) argues. Thus, typically, there are multiple leaders in a group.

In a valuable early insight, Goodall (1965*a*:454) reports that through close study of the interrelationships of the first apes to visit her camp regularly (the charismatic David Greybeard and his dependent companions William and Goliath), she realized that, while a particular individual might often appear to be a 'dominant' leader when with some other particular ape, 'its dominance status with regard to other individuals may change according to the situation.' I suggest that among chimpanzees the true (charismatic, male or female) leaders' main concern within the role is *maintenance* of the loosely structured group of individuals whose goals are primary. i.e. food and reproduction. Individuals alternate between leader and follower roles not as group goals change but as *situational needs* change.

Nishida's and de Waal's insights

Nishida (1979) and de Waal (1982), both of whom subscribe to the theory of a male dominance hierarchy as the basis of social organization of chimpanzees, recognize that aggression-based dominance is not the only form of leadership among the stressed (artifically fed Mahale, and captive zoo) groups with which they are so familiar. This is a most important insight.

It is mentioned in Part 1 that de Waal (1982:211) recognizes that coexisting with the 'real dominance' of alpha males won and held through aggressive competition against other members of the group, there is also a second, far less easily discernible, form of social control based on 'a network of *positions of influence*' (my emphasis). De Waal (1982:211) points out that as long as we concentrate only on the 'real' dominance hierarchy and ignore the importance of positions of influence, our understandings of the social processes of chimpanzee society 'will be very poor indeed.'

Nishida (1979) uses different terms but he, too, identifies a form of leadership based on influence, along with the hierarchical dominance form. He distinguishes between 'alpha' (dominant) and 'top-ranking' males in the Mahale group. A top-ranking male is 'something more' than an alpha male who is dominant over all other males through aggressive contest, Nishida (1979:93) submits. The top-ranking apes do not seek leadership, but, far more frequently than alpha males, they are chosen as the object of following by the other adult males. Nishida (1979:94) mentions that a seasonal group migration which was characteristic of the Mahale K group was due to the 'nonintentional leadership of a top-ranking male.' One Mahale male, Kasonta, whom Nishida (1979:95) identifies as top-ranking, was very aggressive, but earlier top-ranking males were far less aggressive than Kasonta, and they too were notable for the 'high frequency with which they were followed by fellow chimpanzees' (Nishida 1979:95).

Even in the aggressive Gombe society, the ape David Greybeard remained 'an extremely nonaggressive animal', yet he was attacked less often than any other chimpanzee, except for Mike, when he was 'alpha' (Goodall 1968*b*:339). Goodall also reports that David was groomed more by others, and also groomed others more, than any other male. Because the high-ranking David was not alpha (i.e. not top of the aggression-based dominance hierarchy), Goodall (1968*a*:265) reasons that grooming 'cannot be correlated with an individual's position in the dominance scale,' for David appeared to occupy a position somewhere around the middle of the dominance hierarchy. He was, however, by Nishida's criterion, top-ranking, an individual in a position of influence, in anthropological terms a charismatic leader. In Gibb's terms, these top-ranking animals are 'true leaders', as they meet his criteria of having been chosen by their followers. Alpha (aggressively dominant) apes are not leaders, but self-imposed despotic 'heads' of the group.

Both Nishida and de Waal also report that top-ranking or influential chimpanzees groom and are groomed far more than are the alpha animals, and they also have far more opportunities to copulate. I suggest that these observers have not only identified charismatic leadership still existing in these disturbed (fed or captive) groups, but they have also supplied important means of identifying charismatic leaders. Charismatic members, particularly the extraordinarily charismatic, are followed, groomed and copulated with far more than others. The social significance of this preference is explored in Part 6.

In the egalitarian human foraging societies, charismatic leaders, persons with influence, are both male and female (Turnbull 1968*a*; Lee 1979). This fact, which is important to our true understanding, is acknowledged but underemphasized in the anthropological publications. The evidence indicates that the same principle holds among chimpanzees. Charismatic leaders who attract followers are of both sexes (see In search of the female, Part 6).

Traits of charismatic leaders

Lee (1979) reports that in human foraging groups no single personality type or trait is characteristic of all leaders. He suggests that what foragers' leaders share is an absence of certain traits: 'none is arrogant, overbearing, boastful, aloof' or acquisitive (Lee 1979:345). Such traits 'absolutely disqualify a person as leader and may engender even stronger forms of ostracism' (Lee 1979:345). Immediate-return human foragers do not value prestige (i.e. a high reputation arising from success of some kind). What these foragers value most is peace between individuals and groups, autonomy, self-esteem and generalized attachment to social, as well as actual, kin.

Turnbull (1965:304) emphasizes that authority 'is divided among the entire population of any . . . [foraging] band' and that even his attempt to describe the nature of authority 'perhaps suggests too much systemization, for there is so much overlapping.' The leadership role seems not sought after nor especially desired in human foraging groups and it is shifting and temporary (Service 1966; Turnbull 1968*a*). In such societies the 'followership' role – which is a much underexamined and equally vital part of the social structure – also is shifting and temporary. (That the charismatic and dependent status/roles are normally dynamic and situation-dependent are points that cannot be overemphasized). A man or a woman might be assigned the leader status/role (that is, have influence on the behavior of others) because of some skill, or knowledge of some ritual, but that same individual will slip easily into the follower role if circumstances change, and become such that reliance of the group on some other individual is advantageous, through his or her possessing some other skill or knowledge more useful under the new situation.

To turn again to the chimpanzees, in a sense, each time any chimpanzee beyond the age of childhood dependency encounters another there is a decision to be made as to which status/role each individual will assume. Individual adult chimpanzees more than humans constantly change from follower to leader status/roles and vice versa, according to circumstance. A confident young adult might have attracted some less confident members to follow his lead, but on encountering a more charismatic – perhaps a more experienced, older – chimpanzee, the youthful leader in turn might be attracted to follow this individual and so take on the follower status/role. The same ape who for a few hours followed a charismatic member to whom he or she was attracted might on meeting another subgroup be attracted to join that group. The new member might be more influential on the basis of requirements of the situation or qualities of character than the animal leading. In this case, the newcomer will find itself now being followed, in contrast to its role in the previous subgroup.[23] These decisions are not difficult *because the relationship is not a competitive one*. The essential egalitarianism makes these changes easy. Both status-roles are equally and mutually necessary.

The nature of the follower role

Unfortunately, there is a dearth of both evidence and theory regarding followers, human or chimpanzee, which blurs our understanding of the balance of power that exists in egalitarian societies. Concentration of the observers' interest is on the aggressively dominant, and the outstanding 'personalities'. Little is reported of the less impressive, 'average' chimpanzees.

It is commonly assumed that followership is connected with low rank and submissive behavior. Followers are thought of as playing a less important role in maintaining social order than do dominant individuals, but research into the nature of small human groups indicates that followers are active, rather than passive, group members, and that group and leader function depends heavily upon them (Gibb 1970*a*).

It has been established that according to leadership theory, being a follower – or a true leader – is not a fixed state. Nor do (human) followers accept as leader someone 'utterly different' from themselves (Hollander 1970:294). They choose a member of their own group whom they perceive as having superior task competence required in the current circumstances, who may, when situational needs change, be prepared to follow, to take the dependent role. Being a follower is not inconsistent with being a leader at another time, Hollander (1970) points out.

A sense of dependence to varying degrees is found in all human beings (Nisbet and Perrin 1977). Possession of a fluid tendency toward both positive, normal charisma and normal dependence by each average individual makes easy the shifting from one status/role to the other that is characteristic of the mutual dependence system.

Silberbauer's (1981:462) experience in studying the organization of the Bushman (San) of the central Kalahari Desert was between 1958 and 1966, before their traditional foraging way of life was disrupted. According to Silberbauer (1981:462), at that time, each band was a 'consensual polity', and leadership a fluid and ephemeral role which served 'only to guide the band toward the consensus that is the real locus of decision.' Consequently, an individual's or a family's autonomy and self-reliance was not diminished by dependence on the authority of the leaders. Dependency in these foraging groups is a positive form, connected with the expectation and subjective experience of being helped and reassured, not subjugated.

Among both undisturbed chimpanzees and human foragers living traditionally dependence is as Bowlby defines it: a matter of trustful reliance upon others, without implication of any 'pejorative flavor' or negative quality such as helplessness or subjugation. Bowlby (1973) suggests that assumption of the dependent (follower) role hinges on recognition of the higher competence of another in a particular situation, and a trusting willingness to be guided by the more competent partner, while the situation lasts. 'Dependence and independence are inevitably conceived as being mutually exclusive,' Bowlby (1973:37) points out, but 'reliance on others and self-reliance are not only compatible but complementary to one another.'

Bowlby (1973) comments that trusted persons or 'attachment figures' (Bowlby 1969) who provide a secure personal base are required by mature adults as well as young children, although not so obviously, or as urgently. Because of our Western values, the requirements of adults for a secure base tends to be overlooked or even denigrated, Bowlby (1973) suggests. On the basis of his studies of both human and nonhuman primates, Bowlby (1973:24–5) deduces that the essential ingredients of a healthy, self-reliant personality are a capacity to rely trustingly on others at times, and to be able to exchange roles when the situation changes, and in turn to provide a secure base from which his or her companion can operate, at other times.[24]

The nature of deference

Macaques, baboons and apes are primate species which accord privileges and respect to aged individuals (Jolly 1972). Rowell (1972:180) suggests that there must be something comparable to (human) deference among nonhuman primates, if the group is to draw on and profit from the experience and long memories of elders, 'in such situations as choice of feeding route or predator avoidance in unusual circumstances.' Shafton (1976) points out that this deference would not occur if aggressive success were the only means to high status.

Kortlandt (1962:130–1) was clearly impressed with the deference paid to the 'grand old man' in his study group, a very elderly, bent, slow-moving chimpanzee, who seldom participated in the male displays, yet whose 'whims and fancies' were indulged and whose company was sought after by even the prime adult males. Kortlandt's 'grand old man' could not hold the respect of the prime males through aggressive dominance. Without doubt, this elderly ape was a senior charismatic leader, in anthropological terms, an 'elder'. (According to anthropological understanding, old age alone does not endow an individual as an elder).

Writing on the nature of deference and demeanour in human society, Goffman (1956) suggests that deference is that component of activity which functions as a symbolic means by which appreciation of the recipient's superior skill, judgment, knowledge (etc.) is conveyed to that recipient. 'These marks of devotion represent ways in which an actor celebrates and confirms his relation to a recipient' (Goffman 1956:477). Such ceremonial activities are status rituals, rather than being expressions of rank. They may take either a (negative) avoidance or a (positive) deference form.

According to Goffman (1956), deference is not one-directional, but a two-way exchange. There are also 'deference obligations that superordinates owe their subordinates' (Goffman 1956:479). Priests everywhere seem to be obliged to respond to offerings with an equivalent of 'Bless you, my son', he points out.

A deferred-to person must earn and deserve deference; it cannot be self-awarded, but must be sought from others. In seeking deference from others, the seeker 'has added reason for seeking them out, and in turn society is given added assurance that its members will enter into interaction and relationships with one another' (Goffman 1956:478). Charismatic chimpanzees also must earn and deserve the leadership role, and the reward of deference from those who choose to follow.

I suggest that in the peaceful wild chimpanzee societies it is usually customary and expected deference being expressed, when a younger or less assured animal routinely steps out of the path of a more confident or senior chimpanzee. (Less often is it fear). Such yielding, voluntarily conceding something under the pressure of habit and expected behavior, expresses respectful regard and in doing so accords the recipient a very positive form of dominance. It does not imply or involve a reluctant or helpless inability to resist being subjected to some kind of treatment, i.e. submission, as at Gombe in recent years. Such coerced submission does not signify real respect.

Goffman (1956) refers to positive expressions of deference as 'presentational rituals.' Presentational rituals among humans – bowing, saluting (greeting), hand-shaking (touching) and such (all gestures the chimpanzees use) – encompass 'acts through which the individual makes specific attestations to recipients concerning how he regards them and how he will treat them in the on-coming interaction' (Goffman 1956:485). The regard in which the actor holds the recipient need not be one of respectful awe' – it might be that of 'affection and belongingness' (Goffman 1956:479).

Goodall (1971*a*) and Ghiglieri (1988) describe virtually the same greeting exchange by chimpanzees. In each case a passing female paused and proffered her hand to a male, who responded. The Gombe male patted the female briefly, while the Kibale male 'almost absentmindedly' touched the offered hand (Ghiglieri 1988:168). Such seemingly ritualized exchange fulfills Goffman's (1956) model of leader–follower presentational rites. It is not by definition a male–female exchange.

Because of the autonomy and self-direction of individuals which leads to frequent change of companions and pursuit, and the mutuality of the charismatic and dependent status/roles, the situation between participants in a charismatic–dependent relationship is one of low power. In such a situation both actors must exercise great care in their interactions with those on whom the relationship is dependent, because either can opt out of the relationship at will (Thibaut and Kelley 1959).

The charismatic–dependent role relationship

A generalization of the parent–child relationship

In human foraging societies authority based on power is strongly resisted. Service (1966:82) suggests that the only authority accepted is that of parents over their children. More accurately, I suggest, this is the only *form of authority* accepted by foragers who live by the immediate-return system, including undisturbed chimpanzees. Montagu (1966) pointed out that social (or cooperative) behavior is simply a continuation of the parent–offspring relationship. As Midgley (1978:136) puts it, in societies of higher animals 'wider sociality' in its original essence is the ability and inclination of adults 'to treat one another, mutually, as honorary parents and children.' These quasi-parental bonds work well, Midgley continues, 'because they are adapted to soothe, . . . to forge a bond. Once forged, why should this bond not carry its usual consequence of protectiveness?' After all, a fundamental element of parental care is defense and rescue from danger, Midgley reminds us.

To go back to Russell and Russell's (1972) suggestion that the male and female parental roles of nonhuman primates are different, but complementary, it may be recalled that these scientists' hypothesis is that the foremost male parental role is one of *generalized* protection of all of the dependent young of the group, while the foremost female parental role is intimate and *personalized*: a concentration on nurturing her own biological child. There are multiple 'social fathers' and individual 'biological mothers.' Even earlier, in a 1964 publication, the Russells suggest that the *leader–follower relationship among undisturbed primates generally is essentially that of the parent–child, generalized toward the whole society*. Little attention has been paid to this extremely important insight by those interested in the social behavior of chimpanzees, yet the charismatic–dependent relationship of this species takes this form, of guiding, protective (charismatic) 'parents' and trusting accepting (dependent) 'children.'[25] The same supportive, generalized parental form of authority also organizes the highly independent human foragers.

There is considerable empirical evidence of chimpanzees acting in a fashion which suggests such generalized parental authority. For example, Albrecht and Dunnett (1971:32) write of an occasion on which a confident adult female, Pandora, and a youthful male began striking out at each other. An elderly ape, MacTavish (who was usually extremely nonaggressive), did not hesitate in running at the pair and slapping them both, which stopped the squabble. MacTavish's quick response, and the acceptance of the blows and ending of the undesirable behavior on the part of the stronger, younger apes suggest that the older ape had some form of authority (beyond influence) over them. Goodall (1971*a*) and de Waal (1982) report virtually the same type of interaction. Goodall (1971*a*:133) writes of an incident in which two juveniles were squabbling over a banana, and an elderly male ran toward the pair and 'cuffed' both of them, stopping the quarrel. One possible reason for adult apes accepting such intervention by elderly members is offered in Part 6 (see 'The role of charismatic elders').

The ape David Greybeard had quickly developed amazing trust in Goodall, allowing her to touch and even to groom him. Later, when she learned field research methods, Goodall (1971*a*) recognized the methodological error of this interference; however, she could not bring herself to regret it – such was the appeal of David's personality. Goodall (1971*a*:259) writes that David fully accepted her as a sort of 'strange white ape'. (In fact, in the early years, the confident old female Flo also accepted Goodall in the same way, trusting her near her infant children).

Goodall relates that in the early days she spent many days alone with David, following him through the forests for hours, watching while he fed or rested, struggling to keep up through the tangled vines. Sometimes, she was sure that David waited for her, 'just as he would wait for Goliath or William.'[26] 'For when I emerged . . . [from the thicket] I often found him sitting, looking back in my direction; when I had appeared, he got up and plodded on again' (Goodall 1971*a*:259). He would wait – just as he would for his most constant, dependent chimpanzee companions.

Watching and accompanying a charismatic individual are two prominent aspects of dependent role behavior, and in the very early years Goodall spent a great deal of time watching and following David. It appears, from his behavior and attitude towards her, that the confident male does seem to have accepted the nonthreatening and attentive 'white ape' as yet one more of his dependent followers.

MacKinnon, too, had this experience. When he worked out of the Gombe Research Center in 1965–6, carrying out entomological studies, he often encountered some of the habituated Gombe apes in the forest. He found that most of the apes were shy of being followed, but one young adult, Pepe, who was 'least shy of all,' was not disturbed by McKinnon following him. 'Far from trying to lose me as other chimps did, Pepe would sometimes glance over his shoulder and wait for me to catch up' (MacKinnon 1978:69). This young male 'moved easily through the various circles of the chimpanzee community in a manner quite distinct from anything we have seen in other apes . . . but very similar to social complexities common in human communities' (MacKinnon 1978:76).

In commenting on the relationship that grew between herself and the ape David Greybeard, Goodall (1967*b*:191) writes that her close contact with the charismatic David led to the establishment of a bond between them – a bond based on '*mutual trust and respect* – in a sense, a *friendship*' (my emphasis). Their communication was through the emotions, and the message that the human received was one of trust, reassurance and acceptance.

I fully accept Goodall's assessment of this remarkable cross-specific friendship, mutually communicated. However, this relationship is reported here as an example of an interaction between a charismatic chimpanzee and an individual (albeit human) whom the charismatic ape accepted and responded to as to any dependent companion. Such mutual trust, respect, and friendship especially between dependents and charismatics, are structural and functional prerequisites of the mutual dependence system, whether the society is human or chimpanzee.

In order to maintain the basic relationship, the charismatic apes are as dependent on less assured animals seeking their companionship as are the dependents upon obtaining it. Such mutuality of need for support and companionship operates to maintain and balance the two status/roles. The constant change of most actors from one to the other role acts as a leveling device, keeping individuals egalitarian. It also keeps the status/roles of leader and follower from becoming two separate social strata, classes or ranks. At the same time, the constant change of companions and roles reinforces the positive, affiliative social bond between the society as a whole, and keeps gangs, cliques or permanent coalitions from forming. In the highly positive, ever-fluctuating, (wild) chimpanzee social groups, each of which contains autonomous chimpanzees who are charismatic to varying degrees, according to situational needs, there is no way in which a confident animal can coerce a timid one into accepting any act or demand. The threatened ape can simply go off with other, more compatible companions and avoid the coercing animal. Thus, there is a counterbalance between the need of the charismatic leader to attract back-up followers, who will support his or her authority when needed, and the inclination of the less confident to seek the reassuring and protective company of a charismatic comrade who will, if necessary, act on his or her behalf. Through such balances a dynamic equilibrium is achieved and maintained, facilitating the smooth functioning of society, despite the superficial appearance of disarray. Expressed habitually throughout their daily lives, the exchange of deference and response is socially cohesive, role reinforcing and rewarding to both participants – suggesting that the relationship is one of 'non-interference mutualism,' which will be discussed next. The two interdependent status/roles carry different obligations, but yield similar social rewards, i.e. the esteem and affection of others, important elements in the building toward the ultimate goal of self-esteem.

Noninterference mutualism

Wrangham is interested in the question of why, in many species, social groups tend to be composed of particular combinations of members who prefer to associate and interact together. He views this preferential tendency in terms of mutualism, which he defines broadly as being those relationships 'by which two or more individuals gain greater reproductive potential (i.e. expected number of future offspring) than they would by acting alone' (Wrangham 1982:270). All partners benefit as a direct result of their joint effort, as when two or more animals hunt together and share the catch, or keep each other warm.

Wrangham argues that there are two types of mutualism, with differing advantages. One type he refers to as 'interference mutualism' or IM (Wrangham 1982:272). In IM some individuals fare better than others, solely as a result of excluding a conspecific. The other form of mutualism that Wrangham (1982:272) proposes is 'non-interference mutualism' or NIM. This form of relationship would occur if the species consisted of pairs, all of which behaved mutualistically, Wrangham hypothesizes. In the NIM form of mutualism, the actors involved gain an increase in reproductive potiential through their cooperative effort, without reducing the chances, or at the expense, of other individuals. 'Provided that others can themselves behave mutualistically, those practicing NIM fare better than they would alone but not necessarily better than others' (Wrangham 1982:272). As there is no extra benefit in fitness from being kin, partner choice in this sense is immaterial. NIM does not result in inequalities of rank, nor does it require closed or aggressive groups.

The consequences of IM are just the opposite. They include inequalities of rank, lower reproductive potential for non-partners, aggressive inter- and intragroup relations and a stable (i.e. closed) group. Wrangham (1982) suggests that the most simple example of interference mutualism is two animals cooperating to exclude a third from access to desired resources such as nest sites, food sources, territories and mates. Because IM partners benefit and others lose, this form of mutualism favors kin partnerships and aggressive groups (Wrangham 1982:274). He suggests that, in the evolution of human behavior, NIM has been a major influence on cooperative behaviors such as hunting, while IM 'may well have been a critical factor influencing the development of inter-group warfare' (Wrangham 1982:285).

There are gaps in our understanding of either form of mutualism, Wrangham points out. In particular, 'we need more explicit models of the causes and consequences of NIM (Wrangham 1982:284). Because of these gaps, in his article Wrangham limits the influence of NIM to a pressure for individuals to associate; but he visualizes (Wrangham 1982:284) 'that more sophisticated predictions could be generated when factors such as the *need for partners with particular abilities* are taken into account' (my emphasis). He further suggests that groups who are organized on a basis of NIM and IM must be analyzed by different methods.

Wrangham's concept of two forms of mutualism is useful toward understanding the social behavior of chimpanzees. In recent years, relationships among the Gombe and Mahale apes apparently take the interference form of mutualism, since their behavior demonstrates the form and consequences of IM which Wrangham suggests. In contrast, the charismatic–dependent relationship characteristic of wild chimpanzees and human foragers takes the NIM form which Wrangham hypothesizes. Indeed, the mutual dependence system developed herein offers an explicit model of NIM. It goes beyond a pressure of individuals to associate, and the proximal benefits go far beyond solely gaining reproductive benefits. Social benefits are high. Although the concept of mutualism is not usually applied to groups as such, I suggest that the entire social organization of the mutual dependence system (as it is among undisturbed chimpanzees) is based on the principle of NIM relationships – between not only charismatic and dependent individuals, but also males and females in connection with their parental roles, mobile and sedentary subgroups, and even between local units of the larger society through their carnival meetings (see also Part 6, 'The social function of the 'carnival' reunions').

The means of social control

The independent conformists

Sociologists hold that the most fundamental form of social control depends on the individual's acceptance of the standards of behavior defined by social norms and roles as being expected and proper. Accordingly, internalization of social norms and role-related behaviors through socialization is an essential source of positive social control; that is, social control that depends on the positive motivation of the individual to conform.Rewards and punishments then serve as reinforcements rather than as the primary sources of motivation (Theodorson and Theodorson 1969:386–7).

On the basis of Thibaut and Kelley's (1959) definition, a social norm is here defined as being a behavioral rule that has been internalized by both members of a dyad or most members of a larger group. As Homans (1950:123) puts it a social norm is 'an idea in the minds of the members of a group . . . specifying what the members . . . are expected to do under given circumstances' (and will be punished for, if they fail to do).

Norms serve as substitutes for the exercise of personal influence, Thibaut and Kelley (1959) indicate. They function to prevent the development of dependencies on specific other individuals, which are the basis for interpersonal power. Norms 'provide a means of controlling behavior without entailing the costs, uncertainties, resistances, conflicts and power losses involved in the unrestricted *ad hoc* use of interpersonal power' (Thibaut and Kelley 1959:147). Norms eliminate problems such as unsatisfactory behaviors, differences of opinion so that – although not deliberately developed as solutions for such problems – they can act to improve the outcomes attained by the members of a group and so increase their interdependence, Thibaut and Kelley continue. They are a positive form of social control.

Milgram (1974) expands on the concept of positive social control. He suggests that one way to understand the leader–follower role relationship is to think in terms of *conformity* (rather than obedience) to the accepted norms of the group. Milgram (1974) points out that obedience and conformity both refer to the abdication of initiative to an external source, but in differing ways. In human groups, obedience 'occurs within an hierarchical structure in which the actor feels that the person above has the right to prescribe behavior. Conformity regulates behavior among those of equal status; obedience links one status to another' (Milgram 1974:114). Conformity is a homogenization of behavior (but not of personality) 'as the influenced person comes to adopt the behavior of peers.' Berelson and Steiner (1964:60–1) point out that, even in the most informal (human) groups, 'the more closely an individual conforms to the accepted norms of the group, the better liked he will be – and the better liked he is, the more closely he tends to conform. Thus, *there is a built-in psychological reward for people who go along with the group*' (my emphasis). Such reward for conformity is an important means of social control that operates within every set and kind of human relationship, Berelson and Steiner continue.

It seems reasonable to assume that, in a loosely organized fission and fusion society composed of self-directed, essentially self-reliant individuals, either human or chimpanzee, rewarded (positively sanctioned) conformity based on social norms would be a very feasible solution to the very real problem of maintaining both individual autonomy and group social control. It would bring about an individually arrived-at, tacit general agreement in the group, which would make easy the conscious consensual decisions which maintain order in human foraging groups.

Knowing of what chimpanzees are conscious is difficult. However, in terms of function, a high level of individual conformity to the social and behavioral norms would bring about concurring responses to conveyed information. When the majority of a group makes the same decision or judgment, their response becomes a spontaneous group response, the practical outcome of a sum total of individual decisions without any formal collective consensual or moralistic decision. For instance, when Goodall (1965*a*) saw several males immediately strike and bite another male who had snatched food (meat) from an elderly ape, the quick same response suggests that each animal viewed such behavior in the same way: as unacceptable. Interestingly, the attacked male did not try to avoid their blows and there were no signs of injury after the attack.

I suggest that conformity to the behavioral norms is a main means of social control among chimpanzees and human foraging groups.

The use of sanctions by chimpanzees

As was explained in Part 2, human foragers consciously use informal but effective negative (punishing) and positive (rewarding) sanctions to discourage and encourage certain behaviors; in this way strengthening the conformity of members to the adapted way of life. Sanctions are never arbitrary or *ad hoc*; they involve an understanding based on knowledge gained through observation or subjective experience that reward or punishment will follow. There must be a concept of breach (or fulfillment) of acceptable behavior, or there cannot be sanctions. The positive sanctions that foragers use are various forms of acceptance and approval, signaled through seeking out, respect, deference, friendship, affectionate behavior and the like. The usual negative sanctions are shunning, ridicule and various forms of ostracism, usually forcing out or driving away, not formal banishment. In the extraordinary case where a member's deviant behavior endangers others, execution by a consensus of the group is a rarely resorted-to, extreme sanction (Brownlee 1943; Lee and DeVore 1976; Woodburn 1982). At another level in these human societies, in which a person is judged for his or her intrinsic character, the most apparent means of social control is through the negative sanction of public disapproval. A transgressor is made aware of having lost the respect of comrades. Ostracism in the simple form of avoidance and nonacceptance is a powerful deterrent, utilized against society-disturbing deviance, in the highly social, peaceful, human foraging societies.[27]

It is Goodall's (1986a:227) opinion that 'group punishment of deviant behavior' through ostracism (or shunning), as practiced in human groups, 'has not yet evolved in a truly sophisticated way in chimpanzee society.' Instances of social rejection or social withdrawal at Gombe usually involve persistent hostility against some individual whose behavior 'did not actually diverge from the group norm' (Goodall 1986*a*:227). Clearly such inappropriate hostile behavior is either 'behavior without a goal' or scapegoating, not the application of a sanction, so we must reconsider the question of the use of sanctions by chimpanzees. It is de Waal's (1982) considered opinion that chimpanzees are entirely aware of the functional effect of their actions. If this is so, they should be capable of applying, and recognizing the application of, sanctions.

Simple shunning of society-threatening deviants, if necessary amounting to virtual exclusion from the group activities, is a passive method of social control; it communicates disapproval of a behavior without engaging in physical (or overt) punishment. Because it does not require physical force, it can be used by individuals against those much stronger or more aggressive than themselves. Such a simple sanction is powerfully effective in human societies organized wholly around positive social relationships. It should be equally effective among chimpanzees, who also are organized through social relationships.

The ease and frequency with which individuals and subgroups fission and fuse – exit, enter, mingle and realign – make it exceedingly difficult to identify clear use of ostracism by either the Gombe or wild chimpanzees. Still, we can make a few reasonable yet tentative assumptions, based on theory and evidence.

Lancaster (1986:215) defines ostracism among non-human primates as being 'socially determined exclusion from the resources and opportunities necessary to successful reproduction.' Such a definition does not necessarily involve driving out. Nor need it involve a moralistic judgment, or result from the collective decision by the group. Barner-Barry (1986) writes of the tacit use of ostracism (toward a very aggressive human child) by young (3½ to 6½ year old) children in a pre-school group. She observed that, quite separately and of their own volition, children in the group simply stopped playing with the bullying, threatening child, and avoided or excluded him from their social activities. He was present but, without formal declaration, isolated. Accordingly, Barner-Barry (1986:282) suggests that ostracism may be the 'spontaneous outgrowth of a number of similar individual decisions by group members.' (So, I suggest, may be the reaping of emotive positive rewards).

The sociological view is that sanctions may be diffused, applied by a single individual as a spontaneous expression of his or her approval or disapproval of some support of, or deviance from, social norms. For example, in foraging societies (as in most societies), persons with violent tempers or unpredictable behavior are avoided (Draper 1978; Howell 1979). If most of the people, quite separately and without reference to others' behavior, take pains to avoid such a person, the deviant is shunned by spontaneous, unorganized, diffused application of the sanction which becomes an affirmation of the social sentiments of the group.

Goodall (1963*b*:297) reports that, because of the large size and uncertain temper of the male Goliath, he was 'well respected' by the other apes, but she illustrates this 'respect' with her observation of Goliath being shunned. When Goliath 'leaps into a tree to join a group . . . the hitherto peaceful chimpanzees scatter in all directions.' In the early years, Goliath was the only male Goodall saw 'actually attacking a female and on one occasion he even drove a young ape from its nest' which he then appropriated (Goodall 1963*b*: 297). The observations on which these data are based are pre-1963 and, at this time, the Gombe group was a functioning wild society. Whether or not the apes were consciously sanctioning Goliath's behavior, or simply fearful of him, spontaneously shunning by retreat, their exit *functions* to ostracize and isolate the deviant.

One might wonder why an independent deviating wild ape might not choose to exit itself, rather than endure the subjective experience of being shunned by many members of the group. Probably often it does. Still, one possible explanation is extremely strong attraction to the group or some of its members.

Marshall writes of the strong desire of the !Kung foragers to avoid both hostility and rejection, which leads these people to conform in high degree to the unspoken social laws. She reports that the !Kung people are 'extremely dependent emotionally on the sense of belonging and on companionships. Separation and loneliness are unendurable to them' (Marshall 1976: 287–8). Most !Kung cannot bear the sense of rejection that even mild disapproval makes them feel, Marshall continues. 'If they do deviate, they usually yield readily to expressed group opinion and reform their ways. They also conform strictly to certain specific useful customs that are instruments for avoiding discord.' Their wanting to belong and be close seems 'actually visible in the way families cluster together in an encampment' and in the way people sit crowded together, 'often touching against their neighbor' (Marshall 1976). Draper (1973) too remarks on the obvious pleasure the !Kung get from sitting in close physical contact with others of the group. Rejection is unendurable to these independent people, so, according to Marshall, simple, friendly acceptance by the group generally carries an emotional reward. We cannot know if such need, and reward, is experienced by chimpanzees in their grooming groups, but it seems possible – even probable.

Human foragers are entirely conscious of what they are doing when they apply sanctions such as ostracism against a socially deviant individual. By shunning, by not choosing to follow, or by not permitting a follower, or by refusing to mate with some particular ape, chimpanzees functionally sanction undesirable behaviors. As de Waal (1982) points out, when an ape intervenes between two quarreling fellows (as Luit did) the act may be conscious but we do not know precisely of what the intervening animal is conscious. Still, it seems reasonable to assume that such behavior is an enactment of the generalized parental role of the charismatic leaders, a conscious intervention to protect the more vulnerable (in that sense, childlike) of the antagonists. At the same time, such intervention functions to restore peace in the group.

In the Arnhem zoo enclosure, the (largely female) group is unable actually to shun aggressive members, such as the young male Nikkie, who for a time became a despot in the zoo group. This aggressive ape was 'greeted' by the other males, but he met 'great resistance' from the females who for some time would not greet him. He was grudgingly obeyed, but 'feared rather than respected' (de Waal 1982:149). Nikkie did not have the qualities of a leader, de Waal explains; he did not keep order and so did not gain the respect and support of the group.

De Waal reports an incident in which the group did appear to sanction negatively the action of the young adult male Luit, who, in the excitement of a dominance-challenging display, used then non-typical physical force against the respected older charismatic Yeroen. According to de Waal (1982:93), the displaying male deliberately hit Yeroen a hard blow and 'pandemonium' immediately erupted among the watching members. Luit was instantly attacked by ten or more chimpanzees and driven as far from the group as possible, to the far side of the zoo enclosure. Luit did not have the normal option of wild chimpanzees, of temporary exit or emigration from the group. It is interesting (but we cannot assume, significant), that a few years later Luit was killed – murdered or executed – by his erstwhile male companions.

It is not clear to which animal Goodall (1968*a*:265) refers in an early report, based on data collected in 1960–5 (when the Gombe apes were categorically wild), when she asserts that the male who was groomed least 'was fairly high ranking and aggressive;' but there may be significance to the fact that despots Goliath and Mike were groomed by each other more than they were by any other ape (Goodall 1968*a*:268). Perhaps these males were genuinely indifferent toward social grooming. On the other hand, they may have known that their actions were such that they would be refused if they requested this important common privilege; we cannot know their motivation. However, as we shall see in Part 6, wide exchange of grooming is of high social significance, and Goliath and Mike were not groomed as much as most apes, by other adult males.

Among human foragers, refusal of marriage is a strong form of ostracism, in a society in which all normal adults marry. Marshall (1976:131) writes of one man among the !Kung, who never hunted, as a 'deviant old bachelor.' He never obtained a wife. Howell (1979:59) writes of a hot-tempered !Kung man in his twenties, who beat his wife. Since that incident 'his reputation for violence has made it difficult for him to find another wife,' and as of 1968 he was one of the oldest unmarried men in the group (Howell 1979:59). It may be recalled that many recorded instances were cited in Part 3, both from pre-1965 Gombe observations and of other wild groups, of refusal of males to groom, and females to mate with certain of the physically mature chimpanzees who solicited these services. It may also be recalled that even after considerable negative change in the social atmosphere surrounding copulation at Gombe, female chimpanzees still sometimes refused the sexual solicitation of aggressive males. At Gombe, in the late 1960s, more confident females still evaded the increasingly coercive solicitations of aggressive males, while more timid estrous females sometimes left the group, alone or in consortship with a male (Goodall 1971*a*). MacKinnon (1978:790) reports that neither despot male, Goliath nor Mike, 'showed much interest in the opposite sex.'[28]

I suggest in Part 6 that any male (or less often, female) not groomed by his or her peers, or not frequently accepted as a sexual partner by a number of the mature apes, has not been accorded full adult rank, no matter what his or her chronological age or physical maturity. If wild females, which are free to choose, will not accept sexual solicitations of aggressive males, if the overaggressive deviant member is incapable of learning to control this tendency, so is disliked or feared and refused mating privileges, *this sexual refusal ensures that the society-threatening individual's genes will not continue in the gene pool*. In proximal or ultimate terms, refusal of mating privileges is indeed a powerful adaptive social sanction. Carried to an extreme, the ultimate result is extinction without killing. I suggest that the usual solution to this extreme form of ostracism among wild chimpanzees is voluntary exit, migration to another group. Nevertheless, unless the behavior of the deviant animal improves in the new group, we can expect that the same negative sanction will be applied.[29]

The point to be emphasized here is that, in terms of function, it is not important that we know whether chimpanzees come consciously to a general agreement or consensus in response to some conveyed information, or whether they as independent conformists respond individually in the same way, thus spontaneously arriving at a group accord, functionally a consensus. The result is the same: maintenance of the social norms.

The positive sanctions: reward and goal

The positive sanctions that reward approved behavior in human foraging and chimpanzee groups are difficult to identify, as the reward is largely a subjective emotive experience of pleasure, joy, the warmth of companionship and the like. Nisbet and Perrin (1977) suggest that in nonacquisitive societies (like those of the human foragers) such intangible rewards are particularly desired and effective. Wild chimpanzees too live in a nonacquisitive society.

Hamburg (1963:305) emphasizes that there is selective value in positive emotional responses such as attachment, affection, respect and love, for both human and nonhuman primates, because such rewarding responses make interpersonal bonds attractive. Writing on the psychology of affiliation among humans, Schachter (1959:2) suggests that people 'in and of themselves, represent goals for one another. They have very real needs for support, approval, friendship, status and so on, which can be satisfied only through relationships with others.' Chimpanzees have such needs. Gratification is both goal and reward. Because in the immediate-return societies of humans and chimpanzees the deepest attachment is to the group, to members in general, these rewards are available through positive interactions and relationships with a wide assortment of respondents.[30]

There is a wealth of communicating gestures, expressions, postures and enactments used by chimpanzees which express positive and negative emotional responses. Some entail postural or facial expression, but in positive response physical contact and many forms of 'approach behavior' involving mutually reassuring touching are frequently used. Patting, embracing, kissing, reassurance mounting, rump turning, and grooming are physical examples that Goodall (1968*a*:282–3) supplies. Goodall suggests that such positive interactions have definite rewarding effects.

Like humans, chimpanzees are self-aware and they share essentially the same psychological constitution. Accordingly, fulfillment of the need for positive affects can be viewed as a powerful reward for approved and needed behaviors, while denial of such need amounts to being a negative sanction. Such actions – or refusal of them – are entirely conscious acts, therefore categorically positive and negative sanctions. I suggest that, among both humans and chimpanzees, being rewarded through a great many positive social contacts can be expected to build self-esteem, a subjective and rewarding experience of worthiness, which mitigates against any adult becoming subservient to any other.

Psychologists consider self-esteem to be the core facet of a healthy human personality. As individuals, we may have personalities composed of different traits, in differing degrees, but these are not important, compared to the importance of the degree to which we are able to esteem ourselves (Buss 1973). The basis on which a person evaluates his or her self is initially self-judged on the behavior and attitude of others toward him or her. The egalitarian nature of foraging social organization and principle of self-direction affords all members a healthy self-esteem. As one !Kung man replied, on Lee's (1979:348) inquiring whether the !Kung had headmen, 'Of course we have headmen! ... In fact, we are all headmen ... each one of us is headman over himself!'

At the conscious level, to know one is needed and valued lends self-esteem to a social being. Through the actions of other apes taking the complementary role, recipient chimpanzees should recognize that they are valued preferred companions. When a charismatic chimpanzee sits down and waits for a follower-companion to join him or her, as both Goodall and MacKinnon observed them to do (see above), or an individual voluntarily follows, is not a value on the companionship of the other being conveyed, which should enhance the recipient's experience of self-worth, i.e. self-esteem? Because of the shifting, temporary nature of the charismatic–dependent alliances, every conforming adult receives positive social contacts from a number of others, a subjective experience of worthiness. Receiving this rewarding response from many mitigates against any adult becoming subservient to any specific individual and deepens their attatchment to the group generally.

The structure of attention

As with lower animals, characteristic stimulus configurations, sounds, body features, postures, etc. are programmed into the chimpanzee's organs of behavior (Shafton 1976). But, far more than lower animals, primates must function in a subtly structured society.

Shafton (1976) argues that natural selection endows each animal with the genetic requirements needed to develop a number of systems of behavior, because having an assortment of systems facilitates an animal's dealing adaptively with particular classes of situation which recur in the life of the species. While capacities for perceptions of unity and community may increase up the phylogenetic scale, they are still features of an animal's inborn behavior system with a present or past survival function for the species, he submits.

One situation that constantly recurs in a group of primates is the need to communicate, to pass information of various sorts among the group. Chance, who first proposed the concept of a structure of attention among primates in 1967, suggests that discerning this structure offers a way of understanding the social organization of primate species 'in terms of the *structure of the system* of communication, rather than solely in terms of the nature of the signal, its content and behavioural effects' (Chance & Larsen 1976:2). A social structure of attention 'is simply a statement of the way attention is organized in a group, within which attention to companions takes place preferentially, as well as to the physical environment' (Chance 1976:315). This infrastructure of social communication is revealed through posture, gesture, facial expression and vocal tone; and differently organized primate species should show differing patterns and social interaction which Chance (1976:324) refers to as 'agonic' and 'hedonic' modes. The agonic mode is one of authoritarianism in which social control is by threat from a dominant or alpha male. Attention of the subservient group is centred on the dominant animal in order to maintain a tolerated spacing from him. This avoidance type of attention structure is characteristic of rigidly rank-ordered primate species, such as baboons and macaque monkeys, Chance (1976:324) suggests.[31] In the hedonic type of attention structure, control is based on positive attraction to leaders. Chance also suggests that their outstanding, nonaggressive display involves mutually supportive relationships and a more relaxed, empathetic social atmosphere. The mode is characteristic of undisturbed chimpanzees.

The nonverbal channels of communication in nonhuman primates act to modulate, as well as to carry, messages, Chance (1975) indicates. He further suggests that, after a researcher has ascertained the structured form of communication of a species, she or he can then consider why some individuals receive information and others do not, what interpretation can be put on a movement or action through the attention paid it, and how and why the receivers respond to these messages in the way that they do.

A distinctive structured form of hedonic attention in chimpanzee groups operates to pass and process information from the most nervous, hence most alert and easily alarmed, chimpanzees, through those whose qualities of temperament are more average, to the most calm and confident members of the group. Some of the latter animals then initiate some action in response to the information received, processed and responded to by the group. At one level the direction of information flow is based on qualities of temperament but, at another level, on the mutualistic charismatic and dependent roles.

It was mentioned earlier than an inborn wide variability of temperament is known to be characteristic of primates. In writing of experiments which helped to establish that individual monkeys of a social group are born with different levels of adrenal response, Rowell (1974) suggests that this variability of adrenal response would be adaptive for animals which live in intensely social groups, in that the members would have a range of responsiveness to potential emergencies. Thus 'there would always be some members which would be alert to apparently minor dangers, and at the same time there would be some animals sufficiently calm to be able to lead a non-panicking group response to more severe emergencies' (Rowell 1974:141). Rowell's is an attractive idea, particularly since it helps to explain selection for the pronounced variability of temperament observed in chimpanzees. Ample evidence has already been cited in this volume which indicates that because of their characteristic confidence the strongly charismatic chimpanzees are less alert to their surroundings than the most easily roused dependents, those whose more nervous temperaments incline them to be sensitive to possible dangers, both from outside the group – from the physical surroundings – and from within.

To borrow Broom's (1981) term, the more nervous apes act as 'alerters'; through their quick overt response to any disturbance or suspected threat, they alert the group. On an occasion when Reynolds (1965*a*) was observing a wild group in Budongo forest, one chimpanzee sighted him and became tense and alert. Reynolds drew back behind cover, but to no avail. The other apes had quickly noticed the animal's tense body and looked towards Reynolds', following the alerter's stare, 'to see what the source of interest was' (Reynolds 1965*a*:42).

Rowell and Olson (1983) suggest that similar monitoring of other's position and behavior, and adjusting of their own in response, is what holds the social group of patas monkeys (also a fissioning and fusing species) together from moment to moment. Hence these scientists suggest that this constant monitoring must have important adaptive significance. Among chimpanzees, the structured form of attention that functions to spread information throughout the group is both an agent of social control, and of cohesion.

There are behaviors that are acceptable among chimpanzees, and others that are not. Deviance that threatens others and, by projection, the society, is negatively sanctioned by the group. Harmlessly deviant behavior among chimpanzees, perhaps such personal eccentricities as preferring unripe fruit, using old, abandoned nests rather than building their own and so on, causes no stir. Such disregard of harmless deviance leaves the group free to 'police' more harmful deviance: aggression, greed, possessiveness, bullying, overroughness with the more vulnerable and so on.

Nervous chimpanzees tend by their nature to be made tense quickly, so when they are relaxed the message conveyed to the group is that all is well. When any disturbance occurs within the group or in the environment, the most nervous animals become uneasy, and this unease is conveyed to others through some of the means mentioned above. Ghiglieri (1988:68) reports that the extremely nervous female Zira reacted with 'exaggerated' nervousness whenever he changed his position even slightly. Even after she became accustomed to his presence, when Ghiglieri moved Zira rushed down from the tree where she and others were feeding and dashed into the forest. But because the other habituated apes did not respond similarly, 'Zira soon climbed back up and resumed feeding' (Ghiglieri 1988:68).

By far the most timid and nervous chimpanzee in the Arnhem zoo was the female Franje. She was always first to raise the alarm by barking when something disturbed her, but since she was alarmed even by a large insect, the others paid little attention to her alarm cries. By contrast, when an alarm was raised by a less nervous adult female or male, it was quickly responded to by the rest of the group, de Waal (1982) reports.

The above reports suggest that the nervous 'alerters' watch the physical surroundings (including what others are doing) and react positively, negatively or neutrally through their posture, gestures and vocalization. Such actions are observed and, in differing ways, responded to – response being, as suggested, a separate decision by each self-directed but conforming individual.

One can visualize that if each confident individual in a gathered large group were to respond to every alarm by a nervous member, the gathering or group could deteriorate into numerous small skirmishes between charismatics (and supporters) protecting preferred companions. For this reason I suggest that there must be a way of processing and sorting information emitted by the nervous members before it reaches the more confident members who will act on it (see Figure 7).

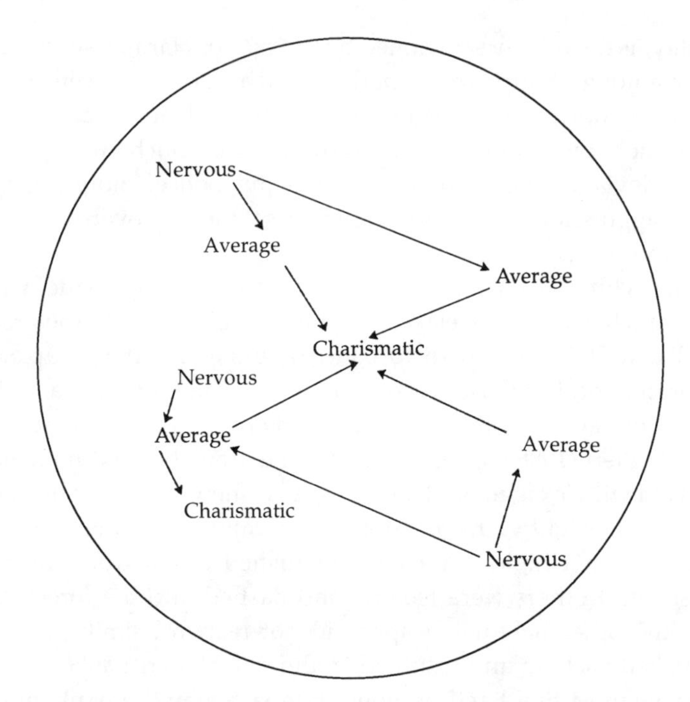

Figure 7. The structure of attention: direction of information flow (a network, not a ladder).

The structure which I deduce entails some one or more of the most nervous members reacting quickly to a supposed or real social or environmental threat, such as unacceptable behavior on the part of another, or even movement in the forest. The alerted animal signals its alarm by vocalizing, physical posture and movement, facial expression and direction of gaze (in Zira's case, by rushing away), and this response will be observed by some others, some of which are less nervous, therefore somewhat less easily disturbed, individuals. The more average apes who (at least in the model) are the majority of the group check the source of the disturbance indicated by the alerters' signals and individually 'judge' the seriousness of the situation, either by responding and echoing the alerters' alarm or by ignoring it. If a large number decide to ignore the alarm, the message is not relayed onward, it aborts right there. The flow of information is broken off without the charismatics becoming aware of the situation. When the alarm spreads through being taken up by many others, i.e. is spontaneously endorsed by the group as being serious, the most confident charismatic members become aware that the group is disturbed and respond to the threat, backed by the group which, by its action, requested it.[32] (In order to fulfill the role on which their status depends, the leaders in a mutual dependence system must rely upon the followers' endorsement, not only to 'back up' their authority, but to initiate their disciplinary and control actions. This social control by conjoint agreement is in harmony with the essential egalitarianism of immediate-return foragers).

Use of this structured form of attention by chimpanzees is not restricted to passing information regarding potentially disturbing or dangerous situations and events and evoking negative sanctions. The same structured response acts to signal opinions, attitudes and individual decisions leading to social acceptance, by the group, of youthful chimpanzees as being of full adult rank and no longer socially children. This usage is expanded on in Part 6.

Among chimpanzees such a structured network of attention functions as a dynamic means of communication and consensual social control. Perceptions of group members are based on reserves of prior information, built up through earlier experiences and socialization, which have become part of their expectations. In a system in which statuses and roles are interchangeable, all adults know what is expected and acceptable through their having experienced both charismatic and dependent status/roles. Every adult has the knowledge to evaluate the performance of any other.

In discussing the infanticide which has taken place in the Gombe and Mahale chimpanzee groups in recent years, Itani points out that an important problem is ignored by most researchers. That is, that characteristically 'there are few counteractions from a third individual or from the group toward the killer' (Itani 1982:367). No attempt is made by onlookers of either sex, other than the mother, to stop the killing. The reason for this apparent indifference or acceptance eludes and worries Itani. Breakdown of the structured form of attention may explain much of the baffling behavior of the provisioned chimpanzees. When direct competition for the withheld and limited bananas began at Gombe in 1965, the focus of individuals' attention shifted. Each ape became selfishly self-centered, interested primarily in his or her own chances of obtaining the desired fruit without regard to what others were doing. I suggest that a breakdown of the structure of attention permitted despots to arise at Gombe, and probably explains the puzzling lack of response to infanticide that concerns Itani.

PART 6

The egalitarian chimpanzees

The next task is to marshal the evidence, largely from naturalistic studies of wild chimpanzee groups, of how the egalitarian mutual dependence system operates within chimpanzee society. Anthropological understandings regarding the workings of human foraging societies are used to structure this discussion.

An egalitarian society is one in which all members are considered to be of equal intrinsic worth and are entitled to equal access to, and share of, the goods, rights and privileges of their society. Within the structure of such societies, individuals have a high degree of autonomy. Hence power relationships do not exist in societies that are egalitarian in form, nor do inequalities of possessions, prestige and status. The immediate-return foraging system is intensely egalitarian. Indeed, Woodburn (1982:432) emphasizes very strongly that no other system found among humans 'permits so great an emphasis on equality.'

People who live as foragers are highly 'individualized;' they are not equal in the sense of being the same as each other (Service 1966:83). The unspecialized nature of foraging societies means that all adults participate much more fully in every aspect of the society than do members of more complex social systems. Members are accepted as they are, for what they are, in their natural variety, Service emphasizes.

Similarities in overt behavioral manifestations of humans and chimpanzees do not necessarily imply that the phenomenon derives from the same underlying mechanisms. However, the following analysis proceeds on the assumption that, if certain positive behaviors and relationships are intrinsic to foraging humans, they are equally imperative to chimpanzees who follow the same foraging system.

A great deal more work will have to be done before we can be sure that the following reinterpretations are correct. But if the reader is persuaded that: (1) this nonhuman primate species is organized around immediate-return foraging as Woodburn (1982) defines it, (2) as Turnbull (1968*a*) asserts, certain positive attributes are necessities without which such foraging societies would collapse, and (3) a pattern of fission and fusion and a dominance rank order work against each other and so are unlikely to coexist as normal components of a social system, then some new understandings of social behavior and organization of undisturbed chimpanzees other than the widely accepted concepts based on the provisioned groups are required.

But first, because it is widely assumed from studies of artificially fed or caged (disturbed) groups that female chimpanzees normally occupy a lower rank than the males, we shall consider the position and influence of female chimpanzees in the egalitarian wild societies.

When field study of chimpanzees began in the 1960s, the dominance rank paradigm was a trusted, widely used model for analysis of chimpanzee behavior and organization. Currently, sociobiological ideas are promoted, to the same end. Use of these theoretical models and over-reliance on the provisioning studies have led to an almost obsessive emphasis on male behavior, evolutionary explanations pertaining to males, and an accompanying unexamined assumption of lesser female status and centrality.

In search of the female

Although studies at Gombe and Mahale National Parks have gone on almost continuously for more than a quarter of a century, there have been no long-term systematic studies of the full social role of female chimpanzees. Riss and Busse (1977:291) point out that the long history of banana feeding 'has obscured attempts to assess the status of females as community members.' Females are usually written of in terms of some aspect of their mothering role or, largely on the basis of (post-1965) Gombe studies, females are written of in terms of the 'copulatory rights' of males to estrous females, as though female chimpanzees are mainly a reproductive resource for the males. This greatly underestimates the normal social position and contribution of the female chimpanzee.

Let us look first at what anthropologists report regarding the position of women in the immediate-return human foraging societies.

The evidence 'certainly does not support a view of women "in the state of nature" as oppressed or dominated by men or as subject to sexual exploitation at the hands of males' (Lee 1979:454). Women foragers, like the men, are usually 'self-assured and self-dependent,' particularly so among the Hadza foragers, Woodburn (1979:256) feels. Generally, women foragers 'give themselves in marriage' (Woodburn 1980:111). Marriage bonds are informal and easily broken by either partner. 'People of influence' – charismatic leaders – are both male and female. Turnbull (1968*a*) emphasizes that, although among human foragers there is a division of labor (men hunt and women gather), these spheres are not rigidly divided. He points out that hunting does not necessarily carry greater prestige than gathering, although meat, being a rare treat rather than a staple food, is greatly enjoyed. (Up to 90% of foragers' diet is vegetable foods, gathered by the women).

In most gathering-hunting societies 'the woman occupies a position of prestige equal to that of the man and is recognized as being equally important' (Turnbull 1968*a*:23). Women foragers have their own spheres of responsibility and control and 'a great deal of autonomy and influence' generally (Draper 1975:78).

Reports by naturalistic observers of wild chimpanzees show that equality of autonomy, influence and status of males and females is characteristic also among undisturbed chimpanzees, as we shall see. Nevertheless, the most detailed examples of how much control and influence female chimpanzees may exert as equal members of the society is supplied by de Waal (1982), as a result of the opportunity for very close continuous study and intimate knowledge of the individual animals afforded through study of a captive group. (Studies of captive apes under superior zoo conditions are valuable, although they illuminate what chimpanzees can do, of what they are capable, not necessarily what they normally do, under wild conditions. But when one has a theoretical framework to work within, it is possible to distinguish which actions are a product of the zoo situation, and which are more likely to be common chimpanzee behavior).

Although, in recent years, the males of the Arnhem zoo colony have generally dominated the females, de Waal (1982:199) reports that 'the social potential of the sexes is less divergent than their actual behavior.' Because, like so many of his peers, de Waal (1982:184) accepts the current view that an aggression-based male hierarchy of dominance is the natural form of organization of chimpanzees and that, consequently, female chimpanzees are 'basically inferior' in rank to the males, he is puzzled by the fact that the Arnhem females often band together to chase off an aggressive or domineering male. He explains the active coalitions among the Arnhem females and the consequent small difference in 'real' (aggression-based) dominance between the zoo males and females by arguing that, in the close quarters of the zoo, females live much closer together than do wild females (de Waal 1982:185). 'In their wild habitat . . . the females live far more scattered,' he continues, so it is extremely rare for females to be able to drive a male away.[33] Being separated, 'wild' females cannot rely on each other for support to control aggressive, threatening males, as can the collected Arnhem females, de Waal suggests. Certainly, the Arnhem females do not have the natural, child-protecting option of exit, as do the wild chimpanzees, which is a better strategy for survival.

When de Waal (1982) comments on the 'small difference' in 'dominance' of males and females evident in the Arnhem colony, suggesting that it is not normal but a product of the special zoo conditions, we must remember that wild chimpanzee females rarely have any need to stop fights or chase aggressive males away. What these incidents do show is that, if the circumstances necessitate it, female chimpanzees, even when mothering, have the ability to act in the generalized protective group role usually taken by the males.

As already emphasized, among human foragers charismatic leadership is both a status and a role, based on influence, not on sex. There are many female chimpanzees described in the publications whose behavior and status indicate that they are charismatic leaders. The elderly female Mama of the Arnhem zoo colony is an outstanding example. De Waal (1982:187) points out that while one male is highest ranking in terms of dominance, in terms of authority through influence, Mama is 'the boss.' Unlike the usually peaceful charismatic in wild groups, in the tense zoo colony Mama was often aggressively authoritarian, yet she retained her influence with others of the group.

De Waal (1982) also describes the leader behavior of a youthful 11 year old female, Amber, who also often acts to protect the vulnerable and to keep the peace. Amber is resolute, of firm character, and already taking the peacekeeping role of the charismatic leaders. Perceptively, de Waal expects that in the future a subgroup will form around Amber. One example de Waal cites, of Amber's peacekeeping intervention, took place when the extremely timid female Franje was holding another female's child, which screamed, alarming its mother.

The mother started towards Franje, then sat down, hair threateningly on end. Franje seemed to be afraid to approach the mother; then Amber intervened, taking the infant from Franje and carrying it back to its mother (de Waal 1982:182).

Although de Waal (1982:210) recognizes the importance of 'positions of influence,' based on 'personality, age, experience and connections,' following the current mode, he, too, focuses on the alpha or top-ranking male as leader over an aggression-based dominance hierarchy. At the same time, de Waal acknowledges that many adult females have authority, which he recognizes when they act to disarm and reconcile male combatants. On many occasions, when two of the males were close to fighting, de Waal saw a female calmly walk up to an antagonist who held a stone or stick and remove it from his hand. The fact that the males did not resist this disarming gave de Waal the impression that the females had some authority to do this. Such mediation 'is a purposeful act,' de Waal (1982:115) submits. I suggest that what this action shows is enactment of the general peacekeeping role of all adults, not specifically authority granted to females.[34]

It was mentioned in Part 2 that anthropologists find it difficult to define the role-related obligations and duties that are part of being a charismatic leader in human foraging groups, because these leaders have no executive power. Fried (1967:82, 88) suggests that such leaders set courses of action which tend to be followed by others and that the action they take is *based on the consensus of the group*. De Waal reports a number of peacekeeping interactions among the Arnhem chimpanzees, in which this form of leadership through informal consensual accord can be recognized. He writes of a typical incident in which Yeroen (a charismatic but aggressive older male) and a young male, Nikkie, chased Luit (a contender in a male power struggle for alpha rank) up a tree, where he sat screaming in extreme panic (de Waal 1982:134). De Waal feared that the outcome of this particular aggressive competition for dominance might be fatal. However, Mama, followed by nearly all of the rest of the group, moved to the vicinity of the trapped male, and sat, hair bristling, staring at Yeroen. When Yeroen and his ally climbed up and began to attack the trapped chimpanzee, the group led by Mama attacked, ending the fight within a minute. In a normal situation in the wild, one or the other antagonist would have probably voluntarily exited, long before the situation deteriorated to this dangerous, society-involving point.

De Waal (1982:55) writes of Mama as the leader of 'collective female power.' But does it not seem more likely that Mama was simply taking the role of active peacekeeping leader, a role usually taken by the childless adults? This was Yeroen's role at the time (de Waal 1982); however, in this situation he was one of the attackers.[35]

There is considerable evidence that shows that wild chimpanzee females also are charismatic leaders. The frail-looking 'dauntless Flo' was clearly a highly charismatic female at Gombe for many years, and the calm, confident old female, Gray, of the Kibale group studied by Ghiglieri (1988) also has the behavioral characteristics of a charismatic leader.

There was a confident charismatic female among one of Albrecht and Dunnett's two study groups in Guinea, whom the observers refer to as 'the bold Pandora' (Albrecht and Dunnett 1971:100). Like Gombe's David Greybeard – and Flo and Gray – Pandora was not afraid of the observers, as were most of the chimpanzees. She was the first to approach the observer's film-hide, albeit uneasily, after which the other apes approached the blind and the observers less apprehensively. Either Pandora or an adolescent male, Scotch – 'not mature ♂♂, as might have been expected' – was usually the first to emerge from the forest onto the observation field, Albrecht and Dunnett (1971:13) report. It should be noted that at their other study site, the ape which habitually came closest to the hide was a mature male. Confident individuals come in both sexes and all ages; personality traits are not limited by sex or age.

These researchers also report that Pandora established a 'number of firsts' in terms of 'curiosity' and 'reaction' (Albrecht and Dunnett 1971:14). They note that it was Pandora rather than the males who was first to attack a dog which entered the observational area, where the group was feeding. On another occasion when a very large troop of baboons entered the field, causing most of the chimpanzees to leave the area, Pandora, who was resting, simply sat up and remained where she was. I suggest that by the criteria established earlier, this confident female is a highly charismatic leader. Interestingly Albrecht and Dunnett tried to identify the dominant animal among the seven adult males of the Guinea group, but were unable to do so because (as is the norm among wild chimpanzees) no animal was clearly dominant.

Some researchers assume that, because at Gombe in recent years the adult females now live alone with their young, apart from the (male) group, this is a normal, pattern of female chimpanzees (see

Wrangham 1979; de Waal 1982:185). The female–female grooming rate is far lower than the male–male grooming rate, and the far-ranging mobile groups appear to be all male. Accordingly it is generally assumed that females are less social and less capable of forming strong bonds than are males.

Wild females are self-directed and independent. Like the males, they sometimes choose to move alone, but usually they are in close proximity to the rest of the group. Nishida (1979) mentions that adult Mahale females sometimes leave the group to give birth, but they usually rejoin the group shortly afterward. Ghiglieri (1984:182) emphasizes that, in the relatively undisturbed (wild) groups that he observed in Kibale Forest, adult females exhibit a very high affinity toward other females. They are '*not* predominantly solitary' like Gombe females in recent years; they frequently groom, and move about foraging, together, Ghiglieri continues. These associating females are somewhat sedentary members of a bisexual society, Ghiglieri argues.

Hypothesizing the general behavior of chimpanzees on the basis of studies of the stressed Gombe apes has led to many misunderstandings. As a result of their Gombe study, psychologists Buirski, Plutchik and Kellerman (1978) report differences of personality in males and females, which they view as being sex-based. They refer to aggressive, competitive, distrustful males and timid, depressed females, a polarization into behavior that they consider as typical of each sex and of the species (Buirski *et al.* 1978:126). I suggest that these are new, localized gender or class – not basic personality – differences, peculiar to the Gombe group since 1965. Qualities of temperament or personality of chimpanzees or humans are not defined by their sex. The essential determining factors in an individual ape's behavior stem from that animal's inner nature, past socialization experiences, role and the immediate situation. Aggression is one condition induced by frustration, but apathy or resignation, 'a giving up and a loss of hope,' is another (Maier 1961:112). There is less danger to infants in the latter.

In an experiment in which a tamed leopard was tethered on the wall enclosing one male and three mothering female chimpanzees, one of the mothers (a wild-born female) and the male attacked the leopard with sticks, the infant-carrying female first and most aggressively (Kortlandt and Kooij 1963:71). Although caution and the avoidance of exposing her offspring to any risk whatever are most conspicuous features of maternal behavior in natural circumstances, this captive female, without the option of exit, did not lack courage or aggressiveness.

Concentration on male behavior, and lack of real consideration of the implications of the parental strategies of both sexes, has resulted in other serious misunderstandings. Nishida (1979:101–2), who is among many who take frequency of grooming between adults as 'an index of the strength of the social bond among individuals,' assumes that more frequent male–female grooming 'may imply that females are egocentric and individualistic, and males are highly sociable.'[36] Nishida's hypothesis ignores the intervening variable of the presence of dependent offspring: that the mother's first task is its care, which includes grooming the not yet competent child. Indeed, grooming is 'the immediate response of the mother to almost all forms of childhood anxiety, fear, or pain' (Goodall 1986*b*:401).

This intimate caretaking attention to his own biological infant is not part of the male's parental role. Being in this limited sense childless, typically the males groom other males. The lesser time spent grooming other adults does not indicate the females' inability to bond with their peers, but a *concentration* of their attention on their dependent young. Such concentration of attention on bonding is highly adaptive when the infant period is a lengthy one of dependence on the mother's care. An affinity for a broader, more generalized form of bonding between males is equally adaptive to the successful survival of the group, and, ultimately, of the species, in that for efficient (generalized parental) protection, often they must cooperate to protect the vulnerable members.

Writing on human same-sex friendships, Hinde (1987:127) suggests that males form generalized (group, wide and numerous) friendships, while females 'are more likely to form close relationships' (fewer, but usually deeper). If we view the bonds between males as *generalized* and those between females as *intimate*, we might better understand our cousin, the chimpanzee, and ourselves.

We must reconsider the quite usual overemphasis on male–male bonds. It is entirely misleading to assign different values ('strong' male and 'weak' female) to same-sex bonding patterns as do Tiger (1969) and Ghiglieri (1989). The evidence indicates that this differential weight is a human assumption, not shared by the noncompetitive, egalitarian wild chimpanzees.

It is a much debated point, whether or not males and females (human or chimpanzee) are different by nature. This is not the place to take up such a controversial issue. But it is clear that we cannot assume anything about the differing natures of the two sexes without keeping very much in the forefront of our thinking the fact that they have evolved differing parental strategies. We are open to grave misunderstanding if we ignore the effects of their primary parental roles on the actions of females and males.

Achievement of adult rank

Although the mutual dependence system is not a rank order, there is a status, and an authority over younger members, connected with adulthood. All adults, male and female, dependent and charismatic, hold the same judgmental power governing acceptance of youthful aspirants to full adult/peer status.

Being accepted as fellow adults, full members of the group, is not merely or firstly a matter of physical size and strength. Indeed, size and strength are relatively unimportant. Rather, acceptance as adult and a peer is based on demonstration of social maturity by the youthful apes, and signals demonstration by each adult member that they accept the youthful female or male aspirant as essentially adult, and no longer merely a large child.

Attitudes toward adolescence, the stage or period of transition between puberty and the attainment of adulthood, differ among human societies and cultures. Adolescence is as much, or more, a product of cultural definition as of biological maturation, and its social and psychological impact on the youthful members varies according to the culture. The transitional stage is long and difficult in some societies and almost missing in others. As we Westerners understand the term, adolescence has come to imply generational conflict, some rebellion of the young against the adults, some exclusion of the youthful, by adults. Anthropologists generally agree that youthful members of undisturbed, traditional human foraging societies suffer little of the uncertainty and confusion that characterize adolescence in the Western world.

Foraging humans do not have the set time and rites of passage from child to adult rank that tribal and some other more structured societies use, or a period of adolescence. There is, however, a recognized period of 'youth' for both sexes, beginning sometime between 7 and 9 years, ending around 11 years of age, during which the increasingly independent youth of either sex is in some things an apprentice, learning from adults (Turnbull 1978; Silberbauer 1981). However, the rights and responsibilities of adulthood are not automatically assumed. As Turnbull (1978) emphasizes, they 'are conferred according to *general opinion* (of the group) as to the ability of the individual' to operate as an adult (my emphasis).

When a !Kung youth marries at about 14 to 16 years of age, he spends the first year hunting as an 'apprentice' to his father-in-law or other mature hunters, until he increases and 'proves' his hunting competence. A young woman must show she is able to gather food and otherwise care for a household (Silberbauer 1972, 1981). The marriage is on a 'somewhat probationary basis' until both spouses show that they are competent, compatible and 'can live in harmony with other members of the band' (Silberbauer 1972:306). The role that the wife's parents play 'is not that of critical examiners, but of helpful and affectionate advisors and teachers' Silberbauer continues.

In human foraging societies there is no attempt to exclude the youth of either sex from moving on from one stage to the next, provided she or he can demonstrate an ability to cope with the additional social expectations and responsibilities. When a boy or girl is ready – self-reliant, competent, confident and responsible enough – to demonstrate the behavior expected of an adult member of the society, he or she increasingly does so, abandoning childish ways. In turn, the adults, finding it no longer necessary to teach or enforce the social rules, gradually begin to accept the aspiring young apprentice as a peer. Through a gradually spreading process, which becomes a general opinion, the group receives the impression that a young individual, through his or her everyday behavior, is acting as an adult.

Most human foragers appear to have no particular formal way of signaling this impression, other than beginning to trust the youthful member with adult responsibilities. Through their doing so, they show that they view that young person as a peer. Among the Mbuti, hunting parties are usually composed of a number of skilled men and several youthful 'apprentices' (Harako 1981). On one such occasion a boy startled the game and was severely scolded for doing so when the party returned to camp. 'Since all members of a band take note of the skills and successes of each hunter' (Harako 1981:530) it was evident to all that the youth was not yet competent in this important adult role.

However, if, for instance, a mature man or woman confers with a youthful girl or boy as to which wild food they should seek and in which direction it might be best to seek it, the adult is signaling his or her acceptance of the young person as a peer, a fellow adult. Others, noticing this acceptance, also begin to think of and act toward the young person as another adult, until that opinion becomes general.

Such spreading attention and appraisal is entirely informal, it does not appear to be organized, nor is it connected with any rite of passage. It is reasonable to suggest that the same network structure of attention operates, as was proposed among chimpanzees. One can imagine that, as the apprentice human becomes increasingly skilled in some area such as healing, story-telling, hunting, gathering or peace-keeping, the adults that are least competent, therefore the most dependent on others in any of these areas, will be the first to be impressed with the youthful member's growing abilities; their acceptance will spread, as suggested above, from the most easily impressed to the more average, to the highly skilled who possesses charisma on the basis of their special skill or skills. For example, the displays by the timid Gombe male William were 'not impressive' (Goodall 1963*b*:294), while those of David Greybeard she describes as magnificent (Goodall 1971*a*). It seems likely that a display by a youthful member would impress William more (and sooner) than it would David.

In her early writings, Goodall (1963*b*) refers to the 'adolescent' period of chimpanzees as beginning at much the same ages as human foragers' children enter the period of youth – between 7 and 11 years of age. While adolescence is a sharply delineated stage among the stressed Gombe apes, and adolescent males must fight their way to acceptance and rank in the hierarchy of male dominance, there does not seem to be anything resembling this period of rebellion and exclusion of youthful males or females in wild chimpanzee groups. There is, however, considerable empirical evidence that suggests that among wild chimpanzees there is a withholding of full acceptance as an adult from the youthful apprentices, to the extent that, in this nonhuman foraging species too, adult rank is not acquired automatically on an individual's reaching a certain size, strength, age, or stage such as puberty.

Among undisturbed chimpanzees, there are both social and maturational factors in the achievement of adulthood. Adult rank is not conceded grudgingly, as at Gombe, on a basis of superior size or aggressive use of strength by resisting adults, hence no energy which might be invested in the vital food search is expended in internal conflict. (A high amount of energy is expended by both adults and young when adolescents must aggressively battle their way to acceptance as a full adult member of the group).

Let us look at the evidence that suggests that youthful wild chimpanzees of both sexes must achieve acceptance as adults and peers through positively impressing on the adult society their ability to fulfill the roles, responsibilities and behaviors expected of adult members of the group. One way youthful humans begin this transition to adult rank is by demonstrating their self-reliance. Goodall (1971*a*) testifies that 'adolescent' chimpanzees act in a more adult manner than they did when they were younger. They make and sleep in their own nest. They play less. They begin to groom others. The youthful male's display becomes impressive enough to draw some attention from other apes (Goodall 1965*a*). From time to time they travel alone through the forest. Even in an open, dry, food-sparse area near Mount Assirik, Baldwin (1979) observed a total of 45 solitary individuals, five of whom she judged to be juvenile chimpanzees and six adolescents (three female and three male). Since Baldwin saw definite indications that the local chimpanzees feared preying lions, such solitary ventures by the young seem to be more than casual. (The anthropologist cannot help but think of the 'walkabout' self-testing of the youthful Australian aboriginal foragers).

Youthful male and female chimpanzees cannot be fully autonomous until they, like the adults, have the ability to be self-sufficient for food. At this age, they are strongly attracted to the far-ranging adult males generally, in part because their attraction to the adult females is not generalized, but largely to their own mother. Consequently, youthful apes of both sexes tend to attach themselves to the wide-ranging mobile groups in which most adult males participate.

Adult chimpanzees are excellent botanists, with detailed spatial memories. They are capable of precise return to food sources by economical routes, from any direction (Wrangham 1977:532). But chimpanzees are not born excellent botanists with an instinctive knowledge of the location of food sources. Which plants are suitable for food, how some (thorny or hard-shelled) are processed, times and conditions of ripeness and locations must be learned by the youthful 'apprentice' apes as they move about with the mobile groups – for they are the up and coming generation of adults who must, in turn, pass this knowledge on. These youthful members are, in Rowell's (1972:180) phrase, 'the valuable memories of the future'.

Through traveling with the mobile group, they also distance themselves from their mothers, thus demonstrating to the group at large that they have outgrown the mother–child dependency bonds. The reassuring presence of experienced male adults leaves the youthful apprentices free to explore encountered situations and to develop their own technical skills. They become known as young adults to encountered groups of 'neighbors' who are likely future mates. As M. R. A. Chance (personal communication) suggests, the youthful 'apprentice' gains a different experience from each adult with which he or she travels. They learn to 'get along' with others who have different personalities. They learn many of each other's idiosyncratic skills, so expanding their own repertoires.

Youthful females, as much as males, must learn these things if they are to be self-sufficient for food and socially mature but, as mentioned, they also have mothering skills to acquire. Therefore they can be expected to travel with the mobile group for a shorter period, and less regularly, than do youthful male chimpanzees, who will become full-time adult members of the mobile groups in due course.

A female chimpanzee reaches puberty at between 6 and 10 years of age and begins to show small swellings of her sex skin, but there is typically a time lag of from 4 months to 2 years after the first estrous period until the female conceives. Drawing on Tutin's work, then in preparation, Pusey (1979:478) suggests that the selective advantage of this period of adolescent sterility, during which female chimpanzees have regular estrous cycles but do not conceive, 'may be that it allows them time to investigate new communities, select suitable mates, and form new relationships.' (Also, adult males often refuse to mate with willing pubescent females. The social significance of this refusal will be taken up shortly).

There is further selective advantage in that it is a period when youthful females are of an age and sufficiently physically and socially mature to join the mobile groups and learn the ecology of their home range. As suggested, the knowledge of what foods are available, when, and where, is necessary in order to maintain the self-reliance for food and freedom of choice in all activities and relationships required by the social system, which are structural rights of adult chimpanzees of both sexes. Without this self-reliance autonomy is lost and individual fitness decreases. It is not because they are in estrus that the female apprentices travel with the mobile groups but because they, like the youthful males, are of a size and strength and social maturity to be able to keep up with and to learn from the food-finders. The young ape, male or female, which chooses not to take this step is still in the juvenile role, whatever its chronological age.

When the inexperienced youthful chimpanzees move about with the knowledgeable mobile adults, the relationship is one of charismatic leader adults and dependent (apprentice) youths, operating at both individual and group level. It seems that the structure of attention also operates in this apprenticeship relationship, in that there is some evidence that suggests that youthful chimpanzees begin by impressing and winning the acceptance of 'lower-ranking' – actually more nervous, dependent – males first, before gaining acceptance by more average members of this group. MacKinnon (1978) describes a typical day in the life of Pepe, a particularly confident youthful Gombe male. For part of the day Pepe traveled alone or with his siblings. Part of the time he traveled with adult males – not the dominant males, MacKinnon (1978:75) comments, but 'low-ranking' adults 'in keeping with his own youthful status.' Nishida (1979:92) too writes of an adolescent male excluded by the Mahale males in 1967 who 'managed to join the cluster of adult males in 1970' by associating 'very closely' with the lowest-ranking adult male.

Somewhat differently Hayaki (1985) observed that the adult Mahale males voluntarily wait for adolescents to follow them. Hayaki cites adolescent males specifically in this 1985 article, but Azuma and Toyoshima (1962) report female adolescents also with the mobile groups in this area, in the early (wild) years. The inclusion of adolescents of both sexes in these groups is highly important in social terms. When the youthful are excluded from membership in these groups, both social and ecological knowledge vital to the survival of future generations is lost.

It may be recalled that, when charismatic qualities and ways of distinguishing both dependent and charismatic apes were discussed, it was mentioned that Goodall (1971*a*) and Sugiyama and Koman (1979), make the same statement: that 'dominant' males received the most grooming. But Goodall (1968*a*:265) also reports that the charismatic David Greybeard – whom she describes as high-ranking but very nonaggressive hence not a dominant male – both groomed and was groomed more than any other member of the Gombe group except Mike, when he was despot. Nishida (1979:94) also reports that a 'high-ranking' (as distinguished from alpha or dominant) male both groomed and was groomed most frequently. When Mike was despot at Gombe, he was very frequently groomed, but most of his time spent grooming and being groomed was exchanged with the former despot (Goliath) whom he had deposed, rather than any other apes, according to Goodall (1971*a*). This may be reconciliation behavior (de Waal 1982). Nevertheless, generally at Gombe mature males with either influence or power are much sought after as grooming partners.

In the pages following, the argument is developed that, in the natural wild groups, one way in which the adult apes signal their recognition of the youthful chimpanzees as young adult peers is first through accepting their sexual invitations and their grooming services and eventually by soliciting these services from the youthful animals. In this way both female and male adults signal their conscious assessment of the youthful apprentice as a fellow adult. It is not unreasonable to suggest that in this nonaggressive way the youthful chimpanzees of both sexes move into and up to full adult rank. As the reports of MacKinnon (1978) and Nishida (1979) suggest, it is probably easier to gain acceptance as a peer first from the least assured (and next, the average) adults, in that they are likely to be more easily impressed than more outstandingly confident 'personalities.' But when senior, strongly charismatic adults accept the youthful aspirants' services and eventually exchange services with them, what is being signaled to the group is the youthful apes' being 'endorsed' as adults and peers by consensus of the entire group. Here, too, the structured form of attention relays information from the nervous, to the average, to the charismatic seniors. Acceptance by the most charismatic members acts as a 'seal of approval', in that the majority of the group has already signaled acceptance.

Unlike the wild young, the youthful (post-1965) Gombe males must force the adults to accept them into their ranks, as fellow adults. The youthful males move up the hierarchy by impressing others with their demonstrated ability to use aggression against them, if they do not submit and acknowledge the adolescents' rank ascendancy – not as an equal, but above them. Each ape, in moving up the hierarchy, forces some other member downward in rank. In the egalitarian social order of wild chimpanzees, however, the status and rank of older members is not affected by the advancement of youthful apprentices to adult rank.

Acceptance through grooming: an adult male–adolescent device

A number of functions of grooming have been suggested by researchers interested in this aspect of chimpanzee social life. Social grooming is a hygienic measure (removal of parasites), it soothes, so reduces tension, and it is thought both to indicate and to reinforce social relationships. The following analysis does not contradict these understandings but expands the social explanation.

Goodall (1965*a*:469) observes that during a young ape's adolescence, 'the proportion of time spent in social play slowly decreases while time spent in social grooming . . . increases.' As we know, adult female chimpanzees do not spend as much time grooming adults as do the males. The focus of much of the grooming done by adult females is their own offspring. However, Sugiyama and Koman (1979:335) report that the highest rate of grooming among the wild chimpanzees that they observed in Guinea was held by a youthful female, Pama. Pama spent a great deal of time grooming both mothers of young infants and adult males in whom she seemed to be interested as mating partners.

Adult males, more frequently than mothering females, exchange grooming with their peers, but adult chimpanzees do not groom all of their fellow-apes equally (Goodall 1968*a*; Simpson 1973). Goodall explains that sometimes an ape approaches another and solicits grooming, or it may simply begin grooming the other. Usually the approached animal grooms the instigator in return, but sometimes it does not respond to the solicitation or reciprocate the grooming. Responses vary, depending on the social statuses of the two and their individuality (Goodall 1968*b*:368; and see Table 2).

Youthful chimpanzees of both sexes spend a considerable time in social grooming with their age mates and eventually with mature members. The latter is apparently a privilege: Sugiyama (1969), Ghiglieri (1979) and Goodall (1975:138) report that when a youthful animal first begins to groom an older adult, it is in the capacity of groomer. The adult animal does not, for a considerable period, groom in return. Twenty-seven out of 31 instances of social grooming between adult and subadult males observed by Sugiyama (1969:205) were by the youthful chimpanzees, of full-grown adults. Ghiglieri (1979:233) submits that youthful males and females groom older males 'in fairly equal proportion,' (55% males, 45% females).

Goodall summarizes her observations of 217 exchanges of grooming in 24 months between June 1960 and December 1962, in terms of age and sex (see Table 12 of Goodall 1965*a*:469). A longer and more structured study of social grooming was repeated at Gombe in 1969–70 (Simpson 1973). By 1969, the behavior of the Gombe apes had undergone extensive negative change. Simpson's study has its own high value, but I rely on Goodall's reports of grooming exchanges,

Table 2 *Adolescent–adult grooming; Gombe 1960–2*

Grooming animal	Groomed animal ♂	♀	A♀	A♂	J	I
♂	43	17(9E)	10	11	0	1
♀	15	15(1E)	1	0	12	24
A♀	11	2	9	2	0	0
A♂	14	3(E)	2	5	1	0
J	0	4	0	0	6	0
I	0	1	0	0	0	0

E, Female in estrus; A, adolescent; J, juvenile; I, infant.

This exchange signals acceptance by groomed or grooming adult of youthful chimpanzee as a peer, of adult rank. It functions through the structure of attention. (Adapted from Goodall 1965*a*:469).

since in those early years she was observing much more normal behavior. While Goodall did not take into account the variable availability of grooming partners, still her table does yield an impression, which can be tested in other wild groups. She defines the age and sex classes as infant, juvenile, adolescent males, and females, and mature and old adults (Goodall 1965*a*:432–3). By the end of 1962, Goodall (1965*a*:434) was familiar enough with the local chimpanzees to ascertain that the local group contained 'at least' 15 adult males and 16 adult females. Eight of the females were accompanied by an infant and an older child, one had an older child but no infant. The remaining seven were childless at that time. She also lists numbers of juveniles and adolescents (see Goodall 1965*a*:434, Sex–age ratios) and reports that, excluding temporary visitors, the proportion of males to females seemed to be 'fairly equal' in all age groups.

Goodall's grooming table suggests that wild male chimpanzees groom more with other adult males than with any other category (43 instances of a total of 82 such observations). The category next most favored by males as grooming partners is adolescents, and the males make no distinction on the basis of sex (21 times, 10 female, 11 male). Less often do adult males groom with adult females (17 times) and they do not favor estrous (9 times) over noncycling females (8 times). It is noticeable that adult males rarely groom children (one instance).

Adult females groom their own children (36 times), and adults of both sexes (30 times), also divided quite evenly, 15 males and 15 females. They seldom groom adolescents of either sex (once). There is no mention of their exchanging grooming with juveniles other than their own offspring.

Youthful (adolescent) males mainly groom adult males (14 instances) and less often age mates of either sex (adolescent females 2 times, males 5 times). Adolescent males groomed (estrous) adult females 3 times. Adolescent Gombe males were not seen to groom adult females not in estrus. (However, Hayaki (1985) reports that adolescent males in Mahale National Park groom nonestrous adult females, usually after copulating with them).

Goodall's table also shows that adolescent females, too, groom adult males (11 times) and their age-mates also 11 times (9 times their own sex, 2 times males). They were not seen to groom a juvenile child or infant, and only twice was an adolescent (female) seen grooming a noncycling adult female. In summary, it seems that, other than a mother grooming her own child, *adult apes do not groom children, nor do children groom adult animals*.

It is to be expected that adult female–female grooming will be less than adult male–male grooming. Time spent by a mothering female in grooming her own offspring is time that might be spent in grooming others. It was indicated earlier that, because they groom each other considerably more than do females, male chimpanzees are frequently viewed as possessing a greater capacity for affinity than do the females. However, as indicated above, Ghiglieri (1979, 1984) emphasizes that the affinity between the wild females in Kibale Forest is as at least as high as that between the males. The difference in grooming behavior tells us less about sex-based differences in capabilities for affinity than it does about differing parental roles.

While the general pattern of mutual grooming seems to indicate that grooming is an adult exchange, largely between adults of the same sex and that, in general, adults do not groom children, nor children groom adults, many observers also report initially nonreciprocated grooming, typically by youthful chimpanzees, of adults of both sexes. It seems part of the pattern for adolescents of both sexes to groom adult males considerably more than they are groomed by the males in return.

When youthful chimpanzees begin to join the adult male grooming sessions, they do so at first only in the capacity of groomer. Goodall

(1968*a*:264) reports that there were occasions when adolescent females or males groomed adult males 'for up to thirty minutes' without the mature animal reciprocating at all. Ghiglieri's (1979:233) observations agree with these findings: nonreciprocity among grooming pairs 'occurred primarily when adolescent males and females groomed older males'. Termination of a grooming session usually occurred 'when a groomee failed to switch roles and reciprocate when the groomer stopped and presented to him to be groomed' (Ghiglieri 1979:224). Occasionally the youthful ape would regroom the adult for several minutes longer, then stop and apparently again wait to be groomed. Sometimes the animal being groomed simply got up and walked away. 'It was clear that most groomers wanted to receive grooming after they had groomed,' writes Ghiglieri (1979:233). 'My impression was that the initial groomees knew this.'

Eventually, however, some adult will reciprocate after a youthful ape grooms him or her. Since grooming is usually between adults, when that happens I suggest that the young animal has been accepted – by at least that one adult – as a peer.

But, as we know, reciprocal grooming is sometimes refused: acceptance can be withheld.[37] Goodall (1968*a*:268) writes of one adolescent male who appeared to have 'a marked preference' for grooming Mr Worzle, who 'was never seen to groom the youngster in return.' This adolescent was apparently not being accorded adult rank, even though he sought it from one of the most nervous 'low-ranking' of the males of the post-1965 Gombe hierarchy. Since Ghiglieri's wild apes act in this same manner, refusal of a youthful animal as a grooming partner is not a result of the special conditions brought about by provisioning at Gombe.

If the nervous adult Mr Worzle had responded to the adolescent's grooming by grooming him in return, some other apes, seeing this exchange, in unconscious or conscious imitation might begin to view the young male as perhaps acceptable as a grooming partner, thereby adding their endorsement of the youth as adult. But Mr Worzle did not respond, so the opposite message was conveyed to the group at large.

Grooming is sometimes considered to be an altruistic act, since it is thought to incur 'a finite cost to the groomer without immediate benefit' (Wrangham 1982:283). However, if the youthful groomer is receiving the benefit of being seen by the group as at least being permitted to groom an adult, I suggest that she or he is receiving an immediate social benefit: a first step towards general recognition as an adult member of the group. Even the nonreciprocated grooming is not as one-sided as it appears.

Grooming also benefits the adult recipient of adolescents' grooming, in an important social sense. While the youthful chimpanzee aspirant to adult rank needs the signaled acceptance of the adult members of the group, it is reasonable to assume that her or his solicitation of any adult acts to acknowledge publicly the applicant's recognition of the older animal's authority (through senior (adult) rank and status) to bestow, or to withhold, peer rank. Thus, intentionally or unintentionally, age-based ranks are acknowledged and reaffirmed in the presence of the larger society. The social benefit is mutual.

As we shall see, adults of both sexes signal their acceptance of youthful apes of the opposite sex as of adult rank through accepting or soliciting the youthful members as copulation partners. In addition, adult females, like the males, have a distinctive way of signaling their acceptance of the same-sex youths as peers. They do this by permitting a youthful female to display her ability to carry out the parental role of the adult female by caretaking ('aunting') their infant, for a time. Aunting, the adult–adolescent, female–female device, will be considered next.

Acceptance through aunting: a female–female way

Aunting is a term sometimes used for alloparenting behavior by females. An alloparent is defined as being 'any individual except the mother or another young infant that interacts with an infant under 2 years of age' (Nishida 1983*a*:4). Nishida (1983*a*) studied aunting behavior among the Mahale apes for 12 months in 1975–6 and found that, characteristically, it is carried out by unrelated youthful nulliparous females, who 'often follow' and 'persistently groom' mothering females, with the objective of being permitted to 'babysit' their infants. Unrelated nulliparous females, much more than related nulliparous females (older sisters), initiated or attempted to initiate aunting of young infants. Unrelated youthful females initiated 202 aunting interactions, while older sisters were observed aunting their infant siblings only 53 times. Furthermore, aunting by older siblings of either sex seems to Nishida more in the spirit of play than of assumed or assigned maternal responsibility, while the spirit in which unrelated youthful 'aunts' interact with the infants impressed him as being one of caretaking responsibility.

Aunting as seen at Mahale differs from that reported among the Gombe chimpanzees. At Gombe, in recent years, possessive and cautious mothering apes permit only older siblings to touch their infants, hence almost all aunting at Gombe is directed at infants by their older siblings (Goodall 1971*a*).

Nishida also noted that mothering females discriminate by permitting, or not permitting, individual youthful females to caretake, carry and groom their infant. Aunting, like grooming, is apparently a privilege sometimes withheld. Sometimes when a Mahale mother resisted a youthful female's attempts to take her infant, the young female groomed the mother doggedly, with the result that occasionally but not always she was eventually allowed to borrow the infant for a time.

Borrowed infants are relaxed and content during these aunting sessions but, 'contrary to the relaxed air of the infant, its mother . . . carefully watches the progress of the interactions' (Nishida 1983*a*:25). This attention on the part of the mother is certainly solicitude for her infant, but it also has further social significance.

De Waal (1982:75) writes of the young adolescent female Amber becoming 'completely obsessed' with Mama's newborn infant Moniek. Indeed, all four adolescent females in the zoo colony were 'fascinated' by the infants, he submits. Amber clearly wanted to take care of Mama's infant, but she had to be very patient, de Waal reports, because the child was 15 months of age before Mama allowed the youthful female to handle Moniek. At first, Mama allowed Amber to carry the infant only for very short distances. After several more months of Amber's patient persistence, Mama gradually permitted her to take on much of the infant's daily care. 'Amber became a second mother: an "aunt"' (de Waal 1982:75).

A number of hypotheses have been advanced as to the significance of this aunting behavior on the part of unrelated youthful females. They are, as Nishida (1983*a*), and Hrdy (1981) point out, at a very suitable age to practice mothering. Accordingly, Nishida (1983*a*:30) endorses a 'learn-to-mother' hypothesis regarding the value of aunting. While undoubtedly skill in mothering is acquired through aunting experiences, the idea that this is the basic reason for the eagerness of youthful and even juvenile females to alloparent is weakened by Nishida's own observation that older sisters of an appropriate age tend to play with, rather than aunt, infant siblings; although they, more than unrelated females, are permitted easier, earlier and more frequent access to infants on whom they can practice mothering skills.

Uehara and Nyundo (1983) suggest that some scholars see little of significant direct benefit to the mother and infant in this youthful undertaking. However, Uehara and Nyundo document a case of two young females adopting and caring for a 2 year old infant for 6 days in which he was inadvertently separated from his mother. They suggest that this infant might not have survived had the two 'aunts' not taken the role of substitute mother. In this, they see benefit to infants and adaptive significance to aunting behavior.

Some researchers argue that there is 'status benefit' to the aunt in this relationship, that the youthful female's status is gradually raised through repeated episodes of alloparenting. Nishida (1983*a*) discounts the 'status benefit' hypothesis. He observed that a young nulliparous Mahale female, who was above the lowest-ranking resident mother in dominance rank, had to groom the low-ranking mother 'earnestly' in order to obtain her infant, thereby, Nishida (1983*a*:30) suggests, reversing their ranks. (Nishida is among those who assess dominance rank through amount of grooming received. I suggest that equalization, attainment of adult rank, is the issue here).

Nishida further reasons that if status is earned through aunting, newly immigrant females should seek to perform this behavior, but they do not. He gives as an example one new youthful immigrant who constantly 'followed adult males, especially the alpha male' (Nishida 1983*a*:30). Even when she was close enough to a mother and infant pair for the infant to begin to interact with her by touching or attempting to cling to her, she retreated and gently pushed the infant back to the mother. She seemed, to Nishida, afraid of angering the mother by touching the infant. Nishida was told by Uehara (personal communication cited by Nishida 1983*a*:30) that later, after this youthful female 'was well incorporated into the group, she began to frequently take care of the infant'.

Nishida does not assume that mothering chimpanzees will entrust their infants to any other female, even relative strangers. The refusal of the young female, when a newcomer, to attempt to take the apparently willing infant is probably an indication of her recognition of expected behavior. There is always a conscious decision on the part of the mother as to who may handle her infant.

Although Nishida does not see any benefit to the 'aunts' in alloparenting beyond learning to mother, he does suggest a number of benefits to mothers and infants. The mothers get extra leisure time to rest, feed and groom. There is direct benefit to the infant, Nishida suggests, in that through the aunting care of any other chimpanzees, male or female, the infant ape is learning that (as Turnbull (1978) suggests of the Mbuti infants) he or she can rely for safety and comfort on a number of 'mothers', i.e. the social group. This experience can be expected to build the child's trust in others, and make easier its eventually becoming attached to the group, rather than to specific individuals. Such generalized attachment is imperative for the functioning and cohesion of immediate-return foraging societies (Woodburn 1982).

Despite his own observation that the relationship between older sisters and infant siblings is playful rather than caretaking, Nishida (1983*a*:31) suggests that 'when an infant is cared for by its sister, the mother gains inclusive fitness if the daughter perfects her skill of child care . . . . Thus the alloparental behavior of older offspring . . . accords well with predictions of evolutionary biology (Hrdy 1976)' (Nishida 1983*a*:31). Inclusive fitness does not seem to be the reason for this behavior, in that it is normally unrelated youthful females who aunt.

Youthful females undoubtedly do take care of infants because, as Uehara and Nyundo (1983), and de Waal (1982) suggest, they are fascinated with the babies and enjoy the task. But, at the same time, the young apprentices have the opportunity of displaying their readiness and ability to handle the important female adult role responsibilities. This demonstration may influence the individual mother and ultimately the group to view and respond to the youthful female as a young fellow-adult rather than still being, in social terms, a child. When permitted to aunt, the youthful apprentice is afforded the opportunity to impress observing adults of both sexes with her mastery of adult female (mothering) skills. The gain to the youthful female is one of eventual acceptance to adult ranks. This alloparenting relationship is one of noninterference mutualism (NIM). All participants gain at no cost to any other.

I suggest that aunting is also a female form of display. Those who are interested in the display behavior of chimpanzees use the term display only in reference to the conspicuous physical display typical of the males – the flamboyant noisy demonstrations of physical prowess, running, jumping, charging, brachiating, swaying and dragging of branches, and so on. Accordingly, it seems assumed that among chimpanzees display is a behavior pattern peculiar to males (Goodall

1971*a*; MacKinnon 1978; Bygott 1979; Nishida 1979; de Waal 1982). Yet an important point on which ethologists agree, is that display behavior may be carried out by both sexes, although not necessarily in the same form. One of the functions or roles of displays by both sexes is that displays are important for successful reproduction (McFarland 1981). McFarland is writing in proximal terms, but what is more important for successful reproduction in the ultimate sense, i.e. the successful raising of the young to the age of independence, than both males and females demonstrating their competence to take on the appropriate protective parental role behavior before they are permitted to impregnate, or to become pregnant?

Because of the aggressive tone of the male displays at Gombe, Mahale and in the Arnhem zoo, which in recent years have been mainly intimidation or threat displays, it is often assumed that the main function of displays is to intimidate others, in order to raise a displayer's position in the dominance hierarchy (Goodall 1971*b*; Nishida 1979; de Waal 1982). However, I suggest that this is an extremely limited understanding of the male physical display. The broad objective of display behavior in either sex is to impress on others the displayer's ability and readiness to do a number of things, as the situation and adult roles require. Intimidation is but one of the protective, defensive devices sometimes required. The most important ability that both male and female chimpanzees must demonstrate in their readiness and ability to carry out the fundamental adult roles: i.e. the *generalized, protective,* parental role of the adult male, or the *intimate nurturing* parental role of the child-rearing adult female, which together offer maximum reproductive success through maximum protection for vulnerable offspring. The aunting behavior typical of nulliparous females is undramatic and inconspicuous, in contrast to the attention-catching male display. But, as has been emphasized, *it is part of the female parental (mothering) strategy to avoid attracting attention,* and of the male parental strategy, to do so.

Acceptance through copulation: an opposite-sex way

In addition to adult chimpanzees signaling their acceptance of youthful members of their own sex as outlined, through grooming and aunting privileges, there is also another way that both male and female adults convey to the group their acceptance or nonacceptance of youths of the opposite sex, as socially adult. It is argued next that *adult* males signal their acceptance of youthful females as socially adult first by accepting their postural invitations to copulate and eventually by soliciting sexual access. Similarly, *adult* females signal their acceptance (or nonacceptance) of a young male as adult first by responding to (or not responding to) the male's sexual solicitations, and finally by inviting his sexual services.

Wild chimpanzee society is an egalitarian one, in which leadership and followership are based on complementary status/roles in mutualistic relationship, with a high level of individual choices, not on dominance by the males. Free sexual choice by both sexes, i.e. the freedom of both sexes to solicit and to accept or refuse sexual solicitation, is part of the characteristic mating pattern. Sexual attraction among chimpanzees is not based on age in the sense of younger females being more attractive than older females. Quite the reverse: some older female chimpanzees attract more sexual solicitations than do youthful cycling females. As we know, the elderly female Flo was extremely sexually attractive to all of the Gombe males when she came into estrus.

As a result of his long observation of the sexual activities of the chimpanzees in the Arnhem zoo, de Waal (1982:156) concludes that the apes 'do not live a life of uncontrolled sexual activity;' to the contrary, 'their sexual intercourse is subject to clearly-defined rules.' He indicates that the Arnhem females choose their sexual partners. So do male chimpanzees. There is clear evidence, from several observers, of adult males refusing to mate with soliciting females, particularly the 'adolescent' females. Mating is multiple but not random.

In Arnhem zoo, statistics show that adolescent females have the highest frequency of mating. However, de Waal (1982) emphasizes that this is not because youthful females are the most attractive to the males, but rather that the captive males are possessive of older females and that this rivalry limits the frequency of younger males' access to mature adult females. The youthful males' not being permitted sexual access to willing mature females excludes the young males from an important means of achieving recognition as being of adult rank in the group.

As mentioned earlier, in the wild groups, and in the early years at Gombe, youthful males mated freely with adult females without interference from older males (Kortlandt 1962; Goodall 1963*b*; Reynolds and Reynolds 1965; Sugiyama 1969, 1972; Albrecht and Dunnett 1971; Ghiglieri 1979). When the Gombe society became directly competitive, the mature males excluded youthful males from gaining acceptance to adult rank by interfering when the youthful apes attempted to copulate, just as happened in the Arnhem zoo. Goodall (1971*a*) writes of adolescents 'hunching' in the male sexual soliciting position or otherwise trying to attract the attention of apparently willing estrous females from a distance, and usually being permitted by the males to copulate, after the older males had done so. When adolescents solicited sexual access from Flo, however, the males became very possessive. Sometimes Goodall saw an adolescent 'hunching' at the old female from behind some tree. Flo usually responded and went towards him, but several of the adult males instantly followed. The close 'proximity of his elders effectively nipped the amorous intentions of the adolescent' in the bud and 'he would retreat hastily' to a safer distance. One adolescent, torn between a desire to copulate with the old female and fear of the males, took a few steps toward Flo, glanced at the males, then suddenly went off in a tantrum, throwing rocks and tearing up grass.

When other females are willing and available to the youthful males, why do they get upset at the older males not permitting them sexual access to this old female? Conversely, why do the adult males refuse them the customary access to a willing female which Goodall observed in early years? It seems that through frustration and the stressful, aberrant situation, both the Arnhem zoo and the Gombe males have roused a directly competitive, possessive, usually latent 'spirit' for which wild apes have no need in their smoothly operating egalitarian system and which is normally negatively sanctioned.

We cannot know whether or not the older Gombe males were consciously denying the youthful males entry to adult ranks by not permitting them the recognition that copulation with Flo, a particularly influential elder charismatic, would bestow. They may simply have been being possessive. Nevertheless, if acceptance as a copulatory partner by the most charismatic female of the group does indicate her acceptance of the youthful 'suitor' as an adult, then the possessive attitude of the post-1965 Gombe males does deny the youth full adult rank. From the frustration reaction that Goodall reports, does it not seem that the youthful males (who apparently are permitted to mate with some of the females) are conscious of this? It seems possible that a social desire for acceptance as an adult by the senior charismatic, rather than sexual desire for this particular elder, (and, in the Arnhem zoo, the older females) is the basis of much of the frustration shown by the youthful males.

Adolescents of both sexes apparently prefer to mate with mature animals, rather than with their contemporaries. Hayaki's (1985) observation of copulations by adolescent males of the Mahale M group shows a strong preference of the young males for copulating with fully adult females. Of nine cases of adolescent male copulation which Hayaki describes, seven were with adult females. Two were with adolescent females. In one of these two instances the pair mated seven times. In 19 of 26 instances of postcopulation grooming, the adolescent male groomed the adult female. Seven times the grooming was mutual. Interestingly, the adolescent males were more likely to play with adolescent partners after copulating than to groom them (Hayaki 1985:158). Grooming fellow-adolescents does not carry the social significance attached to grooming adults.

Sugiyama and Koman (1979:335) drew up a table of sexual relationships in which they show mating partners and copulation frequency in a group of wild chimpanzees in Guinea. A mature female, Jire, who had an older infant, came into estrus twice during the 8 months study period. During these estrous periods, she was seen copulating a total of 16 times. Only five of Jire's copulations were with mature adult males. Seven were with the only adolescent male in the group, and four copulations were with males viewed by the observers (but perhaps not by Jire) to be juvenile.

When a young female chimpanzee reaches puberty, she, like the youthful males, is usually eager to mate, particularly with mature males. When Flo's daughter Fifi first came into estrus, she not only 'responded instantly to the slightest sign of sexual interest in any of the males,' but she 'often rushed up and presented before they had even looked in her direction' (Goodall 1971*a*:181). If a male arrived at the feeding station she would often rush over and solicit mating. Not all of the adult males responded to Fifi's solicitations.

Although a female chimpanzee shows 'adolescent swelling' of the sex skin at about 8 to 10 years of age and is mated by older infants and juveniles, she does not attract sexual attention from mature males (Goodall 1983*b*:4). The phenomenon of mature males refusing to respond to adolescent or youthful female sexual solicitation has been reported by many observers (Reynolds 1965*a*; Nishida 1968; Albrecht and Dunnett 1971; Goodall 1971*a*; Ghiglieri 1984). The charismatic Kasonta, who, as mentioned is the most groomed, followed and copulated with of the Mahale males, was involved in 72.7% (40 out of 55) observed instances of sexual intercourse with three 'newcomer' females who joined the group in 1972. Another newcomer, an adolescent female, followed Kasonta continuously for the first 4 months after she joined the group, but neither he nor any other male was seen to copulate with this young female.[38] Few observers offer explanations of this behavior. Ghiglieri (1984:134) suggests that perhaps there is some sign, which the observers cannot distinguish, through which the males can tell that a young female is not fertile, hence not worthy of their sexual attention. In function, this sexual refusal seems to be exactly the same as adult males declining the grooming attentions of 'adolescents,' or mothering females refusing to permit youthful females to caretake their infants. Might not the adult males be declining to mate with a female who is biologically, but not socially mature? Perhaps, since they do not groom children, neither do they mate with them. It seems that, as among human foragers, the youthful of both sexes must mature socially as well as biologically, in order to be accepted as adults. Through their social behavior they must demonstrate that they are functioning as young adults, not merely as older children. It is not only their demonstration of social maturity but also acknowledgment of this maturity by the adult members that conveys adult social rank on the youthful aspirants.

The periodic larger social group 'carnival' gatherings of human foragers and chimpanzees are yet to be discussed. One point, pertinent here, is that although there is a great deal of copulation in the excitement of the larger group gatherings, few pregnancies result from these matings (MacKinnon 1978). Most pregnancies result from consort relations or other calmer sexual encounters (Goodall 1975:157). Because the 'great volume' of sexual activity in the Gombe feeding situation seldom results in pregnancies, MacKinnon (1978:77) suggests that the sex act in these larger gatherings sometimes seems 'no more than a polite way of greeting another chimpanzee'. It may serve 'more of a social than a reproductive function.'

Such social copulation might be viewed as a 'presentational ritual' through which the actor attests to how he or she regards the other (Goffman 1956:485). In this way appreciation of the recipient is expressed. Through such ritualized presentations the recipient is informed that he or she is accepted by, and involved with, the actor, Goffman explains. If, as the evidence suggests, mature adults of both sexes refuse to mate with children, does this not reinforce the suggestion that among chimpanzees copulation sometimes consciously signals acknowledgment of the partner as a peer, a welcomed equal, much as does an exchanged handshake between humans?

It is clear that a youthful female's willingness to copulate is not an automatic 'Open Sesame' to acceptance as a member of the adult ranks. Psychobiologically, the young female is ready to mate, and she does copulate with adolescent and even younger males. But it seems that social acceptance of both youthful males and females to full adult rank is bestowed by the adults. Although there is no direct evidence, it seems likely that general agreement is reached through a structured form of attention, acceptance spreading from the more easily impressed adults to the others, with full acceptance signaled through acceptance by the males and females, outstandingly charismatic, based on the accumulated opinion of the group. Quite aside from the biological function, a youthful ape's requesting copulation with a mature animal may signal his or her feeling ready to assume the social role and rank of the adult members. The other group members, through accepting the youthful solicitor, endorse the young ape as an adult, and so as a member of their peer group.

An interesting aspect of youthful females copulating with mature males reported by de Waal lends weight to the above argument. De Waal (1982:159) reports that although copulation is brief (lasting less than 15 seconds) and the participants' usually expressive faces are 'virtually expressionless, . . . young females sometimes emit a high-pitched shriek' during copulation. De Waal (1982:49) submits that this is a 'special' high pitched scream, but apparently not a scream of alarm or pain or fear. Copulation is not forced on females by any of the Arnhem males.

De Waal (1982) notices that older females do not shriek when copulating, this special vocalization is characteristic of youthful Arnhem females. One youthful female, Oor, who (like Flo's daughter Fifi) was especially eager to mate, more than others tended to shriek when copulating. De Waal (1982) reports that by the time she was 'almost adult' Oor no longer screamed – except when she was copulating with the alpha male. In view of de Waal's testimony regarding the youthful female's continuing to vocalize during copulation with the alpha male, one suspects that there is a functional/social significance to this selective shrieking.

It is now known that far more information is conveyed through primate vocalizations than we previously realized (Steklis 1985). Acoustical variations in vocalization relay different kinds of information about a number of things, including external events and social relationships. Hence, it is quite possible that the youthful female chimpanzee is vocalizing to draw attention to the fact that she has been accepted as adult by one or another of the adult males, particularly when he is an influential charismatic. Perhaps when their adult rank is established and generally accepted, females no longer need to 'advertise' the fact. Whether or not the young female is consciously informing the group, the shrieks do serve to draw attention to what is taking place.

If, among chimpanzees, as among human foragers, adulthood is conferred according to the general opinion of the group, then endorsement through copulation, grooming, or permitted aunting by outstandingly influential chimpanzees such as Arnhem's Yeroen, Gombe's Flo and Mahale's Kasonta would surely carry more societal weight than acceptance by less charismatic adults. In that sense, there may be a benign social hierarchy to climb in order to reach full acceptance to the adult ranks. (Future field observers will be able to verify which adults respond to the solicitations of youthful apes first, and which adults delay giving this sign of acceptance). Such a hierarchy of endorsement would help to explain the 'frenzy' of the Gombe males of all ages, including the adolescents, to mate with the senior charismatic female, Flo, in the tense group gathered at the camp feeding area. Perhaps they were seeking public confirmation, and reassurance of their rank as adults, rather than copulation as such.

The egalitarian immediate-return human groups are not reported to gain adult rank through such exchange. It is the pattern of the youthful applicants having to convince the group, through his or her social behavior and abilities, that they are socially adults and no longer older children that is common to both chimpanzees and human foragers.

The social function of the carnival reunions

It was suggested earlier that the chimpanzee society is an open network of local groups which come together from time to time in carnival-like, highly positive, intensely social gatherings. Chimpanzee carnivals do not usually come about through accidental encounters – although they may do so. They tend to be deliberate meetings, brought about by vocal signals referred to as food-calling. Characteristically, the mobile groups act as the forerunners of their local group, ranging widely throughout their own, and the overlapping neighboring foraging ranges, hooting noisily and drumming resoundingly on tree buttresses as they move through the forest. Reynolds' (1965*a*) impression is that mobile subgroups tend to move separately through the forest in the same general direction, guided and kept loosely together by their calling back and forth. When a subgroup locates an ample source of fruit that is ripe and ready to eat, the vocalizing and drumming increases markedly, attracting other roaming groups and solitary wanderers. Even sedentary groups often respond to the excitement of distant callers, and move to join the congregating group.

Not all meetings of chimpanzee groups result in carnival. Reynolds (1965*a*:158) observed 'many utterly uneventful meetings' of fairly large groups of chimpanzees. Kortlandt (1962) reports that, when the male chimpanzees come out of the forest and join the sedentary group feeding on the edge of the pawpaw plantation, they display. After that initial display and greeting, the chimpanzees tend to ignore each other and to 'go about their business', Kortlandt reports. These appear to be familiar mobile members rejoining the sedentary. Mori (1982) found that, even at the Mahale artificial feeding station, an initial excitement period lasted for about 10 minutes when familiar subgroups of the same local group came together. Then the animals quieted down.

Reynolds also reports observing gatherings in which the excitement continued for several hours, on one occasion through an entire day and the following night. He suggests that members of neighboring local groups also were often attracted by the calling and drumming. In fact, although he was unable to verify it, his impression is that the times when carnivals occurred were 'when groups from two different regions met and joined, either to feed briefly together or to stay on a common feeding ground for a few days' (Reynolds 1965*a*:159).

The Reynoldses observed six of these lengthy, particularly boisterous gatherings. Reynolds (1965*a*) describes the carnival behavior of two groups of about 12 chimpanzees in each, which came together as he and his partner watched. One group was already feeding in *Igeria* trees when the sound of calling and drumming in the forest made them aware that a second group was approaching. When the second group arrived, they leapt into the trees to join the earlier-arrived apes, 'whereupon *all* of them began the wildest screaming and hooting,' running, stamping, swinging, leaping from branch to branch, dashing up and down the trees, occasionally coming close to each other and parting again, and under it all, 'a steady undercurrent of drumming resounded' (Reynolds 1965*a*:158). On this occasion, the excitement lasted for about 55 minutes, after which the apes settled down peacefully to feed. After some time they all moved off in the same direction. The carnival gatherings that Reynolds (1965*a*:159) saw ended 'with each chimp apparently deciding to which group he (or she) wanted to belong and several dashing madly between the parting camps, going their separate ways.' Typically, each animal individually decides with which others it will travel, or whether for a time it will forage alone.

That the larger, longer lasting gatherings in which the excitement is sustained at a very high level for a lengthy period are in fact gatherings of neighboring local groups seems likely. Coming together less frequently, with less familiar individuals, can be expected to lend an excitement to larger society assemblies which would not be roused if the group was a constant unit. Kortlandt's evidence suggests that, like humans, chimpanzees greet others they see very frequently in a much more perfunctory manner than friends and acquaintances they meet with less often. It is these larger, longer lasting, more excited intergroup gatherings that I, like Reynolds, refer to as carnivals.

Goodall saw many such intergroup carnivals in 1960, a summer when wild fruits were particularly abundant in the Gombe area. She reports that she often observed 'excited and noisy reunions' as parties of chimpanzees from the north, and from the south, charged into the valley to feed on the fig crop (Goodall 1986*b*:503). After a period of feeding, during which members of both parties mingled peacefully, the large gathering sometimes divided again, some groups moving off to the north, others to the south. As Goodall did not recognize individuals at that early date, she could not ascertain whether there was any exchange of members; but occasionally both groups moved off together, in a large party, to feed elsewhere. At the time, Goodall continues, she thought she was observing two separate groups 'which frequently encountered each other without undue hostility'. As we know, years later, after the feeding station was established and the same apes began remaining for longer periods in the vicinity, in larger numbers than was common previously, the Gombe observers began to view the feeding station aggregation as one group, which later divided into increasingly hostile northern camp area (Kasakela) and southern split-off (Kahama) groups.[39]

Larger group gatherings continue at Gombe, but they are now restricted to members of the northern provisioned (Kasakela) group. Goodall (1986*b*:162) reports that these gatherings may last a week 'with chimpanzees arriving and leaving at different times.' At these gatherings, 'community members' interact with their fellows, 'play, groom, display, make noise' (Goodall 1986*b*:162). 'In a sense', Goodall continues, the gatherings are 'the hub of chimpanzee social life,' – albeit an attenuated one, in terms of social and adaptive value.

Before further discussion, it should be mentioned that local human foraging groups, too, gather as a larger society for periods of a few days to a few weeks, at a source of abundant water and food. Although the ultimate causes for the recurrent larger society gatherings are probably nutritional and reproductive, still, it does not seem to be the abundance of food, per se, that draws together the larger society. The large human gatherings too, are periods of intense social interaction (Marshall 1965; Woodburn 1972; Lee 1979). Activities such as visiting, feasting, gambling, hxaro (designated partner) trading (gift exchange), marriage brokerage and trance dancing take place (Lee 1979:446).

There is nothing, such as concentration of food or water, to force the extremely independent Hadza foragers together. Nevertheless, Woodburn reports that local groups come together in large camps every 4 or 5 weeks to hold a sacred dance to which great importance is attached. At these gatherings gambling is generally continuous throughout the day. The people are preoccupied with social activities to the point where men are reluctant to hunt; thus, an idea advanced by some scientists, that these gatherings result in improved access to food supplies (through increased numbers of hunters and gatherers), does not apply, Woodburn (1972) argues.

Among chimpanzees, this type of meeting, through summoning other apes when an ample food supply is located, takes place in the rainy season when food is plentiful and easily obtainable throughout the forest. When fruit is scarce or widely scattered, the chimpanzees make little noise and are found foraging singly or dispersed in twos and threes (Reynolds 1965*a*). It is significant that Kortlandt (1962) Bygott (1979) and Wrangham (1977) observe that when the chimpanzees gather as a larger social group, the amount of actual feeding goes down. To Sugiyama (1972:156), the manner in which the chimpanzees feed on fruit and leaves when gathered in such groups seems 'exaggerated,' dilatory, even ritualistic, although social interactions are intensive. Sugiyama (1972) points out that these large group recurrent 'fusions' of chimpanzees facilitate communication within and between local groups, and so strengthen social bonds. Indeed, these large social gatherings of local groups in an atmosphere of carnival are an essential element toward forming and maintaining the positive social atmosphere which permits and encourages neighboring groups to interact as autonomous interbreeding units of a wider society.

Many reasons for the food-calling behavior which brings about chimpanzee carnivals have been offered. Field scientists seem to be in agreement that the excited displaying acts to relieve initial tensions, that the meetings renew and strengthen social bonds between individuals and groups, and that there is an increase in copulation and not a great deal of feeding during these larger group social reunions.

Wrangham (1977) tentatively suggests the food-calling of (Gombe) chimpanzees which summons the group, to be a consequence of reproductive competition by males. Less cautiously Ghiglieri (1984:172) suggests that 'the phenomenon of food calling seems to be an adaptation of males which, because of the component of genetic relatedness necessary for some of its advantages, probably arose through kin selection among males.'

Ghiglieri was unable to establish whether any of the responding apes are related, other than mothers with dependent offspring; however, Wrangham points out that the food-calling brings in roaming male as well as female respondents, therefore as many potential rivals as potential mates. Furthermore, although the rate of copulation rises during large group gatherings, it is reported that, at least at Gombe, these copulations seem to be relatively infertile (MacKinnon 1978). Most pregnancies seem to occur from other mating situations. Consequently, Wrangham (1977:536) concludes that the benefit of calling lies in the advantage that the male confers on 'himself [or his kin] in creating or enlarging a social group.' At the same time he indicates that his hypothesis 'may underestimate the functional importance of the complex social relationships of chimpanzees.' Since chimpanzee females have a limited period of estrus, the larger group carnival gatherings may not function to bring prospective mates together at the right time, but it does give both males and females an opportunity to impress by their behavior; to make friends with, and so attract, others. (Moreover, as these carnival gatherings constantly recur, a new female friend might be in estrus another time!)

Since Wrangham offered his explanation, the Gombe group has closed its ranks and its border, so the food-calling does not actually enlarge the social group, except in the very limited sense of assembling those who are already members. If Wrangham's explanation was understood as benefitting the caller and his *social kin*, i.e. members of other local groups, it would fit within the framework of the mutual dependence model.

If, as Reynolds suggests, participants often change groups during these larger society gatherings, the opportunity of joining with chimpanzees outside of their local (natal) group widens the gene pool. Without considerable outgroup mating, the gene pool of a 'home range' group of the wild chimpanzees (or human foragers) would rapidly become very limited and the group very inbred.[40]

It is not necessary that we establish whether either human foragers or chimpanzees gather in carnival spirit with the conscious objective of strengthening social bonds with a larger society. It seems unlikely that chimpanzees are aware that the carnival gatherings strengthen social bonds with other individuals and groups. They may simply enjoy the excitement and novelty of meeting and mingling with old and new friends and the many forms of socialization. Nevertheless, particularly strong bonds are indispensable structural and functional elements in maintaining the social structure both at the level of the local group and of the larger society of which they are units, owing to the fission and fusion pattern operating at all levels of the social organization. Moreover, the intense socialization that is typical of the larger society carnivals suggests that these animals do subjectively experience a 'connectedness' on a broad basis of generalized attraction and attachment to others, not merely on the narrower basis of consanguinity.

It was Reynolds' (1965*a*) impression that apart from the carnivals, very little that is exciting or dramatic seems to happen in the daily life of wild chimpanzees. He suggests that the excitement of carnival relieves the perhaps too even tenure of the peaceful life. Shafton (1976) reminds us that nothing happening in a group is itself a form of expressive behavior. Inaction (like peace) is not a null condition, but a sign that all is well in the perception or awareness of the group. Still, Reynolds' insight is important, and one which is upheld by affiliation theory. But, as we know, excitement is not always experienced as a positive (pleasurable) emotion. Excitement may be experienced in conjunction with negative emotions, such as fear and anger.

In her 1986 book, Goodall writes of the elements of danger,

excitement and tension connected with the recent territorial patrols of the Gombe apes, as 'spicing' their lives (Crook 1986:23). It may be recalled from the earlier discussion of Gombe patrols that it is mobile groups which patrol the borders of their closed territory, searching for (and attacking) 'stranger' chimpanzees. Unlike the calling, drumming approach of chimpanzees intent on friendly encounter, these patrols are silent, tense, and hostile in intent. As mentioned, the frantic, silent race of these apes toward any sign of nongroup members suggests to Goodall *et al.* (1979:49) that encounters with outsiders are extremely attractive. After the patrollers return to 'home ground,' they explode into vigorous displaying, drumming and vocalizing (Goodall *et al.* 1979:49). We must seriously consider whether the hostile patrols are a maladaptive, negative distortion of seeking of contact with neighboring groups, which is normally expressed in carnival. It is social contact under conditions of positive excitement expressed in carnival, not the negative excitement of danger and hostility, that spices the life of undisturbed chimpanzees.[41]

A strong attraction to coming together as a larger society is adaptive and a necessary structural element in the fissioning and fusing foraging societies. Lorna Marshall, who pioneered study of the !Kung San foragers between 1951 and 1961, writes of the cohering effect of 'trance dances', which are held two or three times a month, or more often when larger groups gather, in which 'men, women, children, visitors – everyone' participates (Marshall 1965:271). These dances are overt expressions of particularly intense social interaction. They draw the larger society together as nothing else does, Marshall reports. Indeed they bring people 'into such union that they become like an organic being' (Marshall 1965:271). Coming together, they celebrate the group.

Sugiyama (1972:155) sees a functional resemblance between the 'frenzied excitement' that spreads throughout the chimpanzee carnival groups and the psychological gratification that human foragers derive from the intense excitement of their trance dance ceremonies. He suggests that when the larger society of chimpanzees gathers, 'the excitement of one ape stimulates excitement in others, and the entire party soon becomes engulfed in frenzied excitement through a feedback process' (Sugiyama 1972:155). Obviously these larger society reunions reaffirm and strengthen a group bond between individuals and local groups whose foraging mode of life is one of scattering, in indirect competition for resources. Through the larger society carnival gatherings, human foragers and chimpanzees widen their social world.

Affiliation theory, which is usually used only in analysis of human groups, seems to be useful toward understanding how such behavior acts to increase the bonds of both human foragers, and chimpanzees (see Power, 1988). For example, the larger society carnival gatherings fit within Goffman's (1971) concept of 'arena-related' rituals. Arena-related rituals are the larger, less frequent meetings between socially supportive individuals who are in contact less often than those who meet through their daily lives. The example Goffman (1971:73) gives of an arena-related ritual is a 'maintenance rite.' Parties to such a relationship may engineer a coming-together if there has not been a recent reason for one to occur. Coming together is important for the well-being of the relationship, Goffman continues, as 'the strength of a bond slowly deteriorates if nothing is done to celebrate it.'

Festinger, Pepitone and Newcomb's (1952) human-based findings regarding the specific needs which only interpersonal contact within a group can satisfy, is also illuminating. These scientists suggest that two classes of social needs must be met, if a group is to be successful. The first class encompasses the needs for approval, respect and support, which are gained through singling out the group member as an individual and by paying attention to her or him in a manner which satisfies these needs, as in the charismatic–dependent relationship.

The second class of needs Festinger *et al.* (1952) propose requires 'de-individualization' or submersion in the group, as in carnival. Action as a group facilitates a reduction of inner restraints against doing things or performing behaviors which are often inhibited when a member is being paid attention by others as an individual. They suggest that, although a group cannot satisfy both types of needs at the same time, it can provide both types of situation on different occasions; those that do so are highly attractive to their members.

The individualized attention evident in the charismatic–dependent relationships and the release of the de-individualized carnivals combines to make the mutual dependence way of life successful. It answers emotional and social needs as well as solving ecological problems. The de-individualized emotive release found in the larger society reunions acts both to draw the separated subgroups back together and to blend local units – so reaffirming and reinforcing the social bonds of the larger society.

Fission – which is basically a foraging strategy, but also socially a way of maintaining autonomy while avoiding aggression – also plays a part in strengthening the social bonds. On a wall that I pass, a bit of folk wisdom in the form of graffiti queries: 'If we don't part, how can we meet?'

The excitement of carnival is dependent on it being an occasional gathering and a celebration, a coming together to celebrate the social bond. Through both components of the fission and fusion pattern a maximum of group cohesion is obtained with a minimum of restriction of individual autonomy.

The function of the mobile and sedentary groups

Understanding the summoning by the mobile callers of the sedentary members in terms of society maintenance alone, underrates the complexity of this subgroup interaction. There are important demographic implications, particularly pertaining to the migratory patterns of local chimpanzee groups. (As used here, the term migration refers only to the habitual pattern of nomadic movement of a local group about and within its home and neighboring ranges).

Because the members of the sedentary groups are not very active physically, these groups may appear to be shelters for those less able to contribute to the foraging way of life. To the contrary, the interaction of mobile and sedentary groups in a characteristic mobility pattern functions to keep the whole local group moving from food source to food source, within the loosely defined bounds of their home range, without depleting any source. The sedentary groups are strongly influential and their role is equally as important as that of the mobile group in the adaptive functioning of the movement pattern.

The first thing we need to understand is that those members who compose a typical sedentary subgroup hold a strong attraction for the far-ranging mobile members. The sedentary groups contain the child-rearing females – mothers of some of the mobile group and, for others, prospective mates – who might come into estrus while the mobile are away. They also contain the young, the infants and children who are attractive to all members of the group. They contain elderly charismatic males and females, who retain their attraction for the mature but younger apes. As already argued, gaining the acceptance of these senior or elder individuals as grooming, aunting, and sexual partners signals full acceptance into the group as an adult and a peer.

The second point to be kept in mind is that the autonomous sedentary members do not always respond to the calling of the mobile groups. Kortlandt (1962:132) reports that the mobile groups in his study area returned every few days to the (perhaps more than usually stationary) sedentary group on the pawpaw plantation. When sedentary members do not respond to the vocal invitations of the mobile group which has located ripe fruit, it seems that they will eventually draw back the venturesome roamers. Woodburn (1972:202) reports that among the Hadza peoples, the initiative for taking decisions about moving 'is in the hands of the men' – i.e. a mobile group. But the women (acting as a sedentary group) can accept their decision or reject it, simply 'by continuing to go about their daily activities and making no preparations to move.' The men accept this veto, Woodburn indicates. He also writes of an elderly widow, Bunga, a respected and conventional woman, who moved camp far less frequently than other Hadza people (Woodburn 1968*b*:105–6). Over a 3 year period, 'a total of 67 adults, more than a quarter of the entire Eastern Hadza adult population' had come to live in Bunga's camp for a time.

It is a point of interest that it was in 1972, the year old Flo died at Gombe, that the southern (Kasaka) group stopped visiting the camp feeding area (Goodall *et al.* 1979:34) and the two communities became quite separate.[42] It was when this elderly female first came to camp in estrus that the rest of the group – with the exceptions of David Greybeard and his followers William and Goliath, who already were regular feeding area visitors – first arrived to feed in Goodall's camp. Flo and her dependent offspring visited the camp feeding area 666 times in 11 months in the period from April 1964 to March 1965 (Goodall 1968*b*:316). While it cannot be proven, it is possible that the constant presence of this charismatic elder was a factor in attracting the mobile members back to the camp area.

The reluctance of the sedentary to move over great distances acts as a sea-anchor, keeping the group to a loosely defined home range. Accordingly, the sedentary groups rather than the carnival gatherings, might be viewed as being the hub around which the society revolves. This function is more evident among chimpanzees than among human foragers.

Checking the condition of local food resources and, if the crop is fully ripe and abundant, calling the sedentary to partake are very efficient in terms of maximizing the sustenance of the group. Keeping the local group circulating around a familiar home range, the mobile occasionally returning to the sedentary subgroups, sometimes being joined by them, rather than moving out into unfamiliar territory, is also an efficient use of a foraging range. If the active mobile groups simply roamed, followed by the slower sedentary, they might not locate sufficient food for all. Food obtained through simply wandering would undoubtedly incur higher cost in terms of energy expended, nutritional returns and safety. Remaining behind, as the sedentary do at a good feeding area, until summoned to another, is beneficial to them in reduced energy costs. The sedentary groups' reluctance to move long distances when local food supplies are sufficient, acts to keep the mobile from expending more energy roaming than is actually required in their role as food-finders. Thus no energy is wasted by either group in futile or unnecessary search, which enhances the prospects of survival of the dependent young, and, ultimately, of the species. In addition, because under normal circumstances the carnivals also attract outsiders, prospective mates, it can be argued that both mobile and sedentary groups gain from the contribution of the other at no cost to other individuals. Hence the principle of noninterference mutualism holds at this subgroup level, as well as in charismatic–dependent alliances of individuals.

At the level of individuals interacting, charismatic leader and dependent follower are fluid interdependent statuses and roles based only partly on personality attributes. As has been emphasized many times, most individuals taking either role shift easily to the other if circumstances require it. The mobile and sedentary groups also shift between the charismatic (leading) and dependent (following) roles. When the sedentary respond to the summons of calling mobile groups and move to join them, they are in the role of a dependent group, following the charismatics. When a sedentary group does not respond, and the mobile return to join them, the roles are reversed.

The role of charismatic elders

Old age is usually venerated in human societies which are either organized around attributes of persons or are familial in form, as are foraging societies. In the elemental foraging societies the problems of living in a given environment are the same for all living generations at any time. As Tanaka (1980) points out, normally there are not the sudden discontinuities common to modern complex societies. The same value system and way of life is maintained, 'so that old people do not get left behind on the far side of a generation gap' (Tanaka 1980:108). Similarly, Service (1966:51) suggests that in the largely unchanging foraging society the old know more about life and the environment than do younger adults, hence it is adaptive to accord them respect and authority, and 'above all, to heed their advice.'

Anthropologists agree that some elderly members of the society gain the special status of 'elders,' having an informal authority through experience and respect over mature but younger adults (Brownlee 1943; Turnbull 1961; Service 1966; Lee 1968; Marshall 1976). (Rowell (1972:179) suggests much the same thing regarding deference to elder monkeys, because of their store of experience).

According to Lee (1979), functioning as an elder calls for something more than chronological seniority alone. Many elderly foragers do not have (or wish to have) leader roles. I suggest that those with the most elder authority have it on the basis of retained charismatic status and role. I also suggest that chimpanzee groups contain elderly apes of both sexes, who function as do human elders.

Elderly human charismatic leaders often act as peacekeepers, as mediators and settlers of disputes, particularly those between adults. Turnbull (1978:219) reports that Mbuti elders of either sex sometimes 'wordlessly, insert themselves physically between two disputants,' making difficult continuation of the quarrel. Lee (1979) reports the same peacemaking behavior among the !Kung. De Waal (1982:124) writes of one of the oldest male chimpanzees in the Arnhem zoo group pushing his hands between two females which were fighting, forcing them apart, then standing between them until they calmed down.[43]

Goodall (1971*a*) reports that at Gombe, when two juveniles began fighting over bananas, an elderly male charged towards the squabbling pair and 'cuffed' them both, ending the dispute. It is essentially the same form of intervention that Albrecht and Dunnett (1971:35) report in Guinea, when the ape they judged 'the oldest of them all' (and, in terms of aggressive behavior, second to the bottom on their scale of dominance), the charismatic McTavish[44] thumped the very assured adult charismatic female Pandora and another female with whom she was squabbling, also ending their quarrel.

Why did 'the bold Pandora' accept the chastisement of the elderly and usually nonaggressive male? It is significant that the peacekeepers appear to be charismatic elders, and that when a mature fellow charismatic (such as Pandora) is disciplined, it will be by a distinctly older charismatic chimpanzee. McTavish would have been a confident, impressive, adult charismatic animal in his prime when Pandora was an impressionable juvenile. At that stage she would have learned to respect and trust the charismatic male, according him the usual age-

and role-based deference of a juvenile toward an adult, particularly one who was a charismatic leader. As she, too, became adult, Pandora did not lose the respect she had had toward those who were adults when she was a juvenile. In the wild societies, elderly, even infirm, chimpanzees do not lose their status – as they do under a system based on strength and aggression. Thus, there is little likelihood that Pandora (who learned respect for the older 'law-enforcer' long before) would resist or resent the punishment, whereas she might do so, if reproached by one of her age group.

After 1965, there was much actual fighting between adult members of the Gombe group, and it was Goodall's (1971*b*:91) subjective experience that the closer the social status of the attacked to the attacker, 'the more likely it is that the former will turn and attack in self-defense.' Interventions by the older charismatic individuals into the disruptive behavior of younger adults as well as juveniles can be seen as a social mechanism to stop potential violence and to maintain the peace essential to a foraging society living by the mutual dependence system.

According to Baldwin and Baldwin (1981:248), one of the variables controlling and modifying animal behavior is the existence of 'models located in the environment.' The outstanding models directing adult chimpanzee society are the elderly charismatic animals, in that they are accorded charismatic status on the basis of the childhood experience of all younger members, even prime, equally charismatic, adults.

I do not suggest that elderly charismatic chimpanzees are consciously recognized as elders, in the same sense as are the charismatic elderly in human groups. Nevertheless, they do have authority over younger adults in the group, based on their having been outstandingly charismatic leaders in their impressive prime, who had attracted – and if necessary disciplined – those now mature individuals when they were children. During their childhood, the adults had learned to expect to be disciplined by as well as to expect affectionate care from, any adult member of the group. They defer to the older charismatic apes as they had done when children.

The role of charismatic elders is that of generalized social parent to all, but most importantly, to the adults of the group. If the childbearing experiences of older females such as the highly charismatic Mama and Flo, are typical, then female chimpanzees continue to bear offspring throughout most of their adult life, hence continue to be involved in the intimate mothering role.[45] Hence, in terms of numbers, most active 'elder' chimpanzees are male. In terms of structure (once again), animals accorded the peacekeeping elder role tend to be highly charismatic, elderly childless apes of both sexes. Forcing the physically less able, elderly chimpanzees to the bottom of a hierarchy based on physical strength and aggression – as happened at Gombe – acts to undercut the basis of adult peer control.

PART 7

Probabilities, possibilities and half-heard whispers

Traditionally, the final section of a scientific monograph is the one in which the author goes beyond interpreting the data to speculate on a range of implications that she or he thinks is recognizable in the data. The following proposals start from a basis of evidence and theory, but go beyond into the realm of logical possibility. The reason for doing so is to stimulate further discussion and to suggest possible new approaches for further research.

The adaptive value of aggressive behavior: a possibility

The theme of this study is that a system of mutual dependence involving positive behavior, egalitarianism, individual autonomy, and the noninterference form of mutualism (NIM) is the adapted mode of organization of chimpanzees under normal environmental circumstances. It is also argued that this benign, loosely structured form of social order of chimpanzees can change rapidly to encompass negative, society-destroying behavior and a rigidly structured dominance hierarchy.

It is now known that equal potentials for aggressive or peaceful responses are both part of the genetic constitution of all animals; there is no genetic coding that inevitably results in either aggression or peacefulness, only a set of genetic attributes ultimately for self-defense 'that can be expressed as aggressiveness under particular sets of conditions' (Dubos 1973:86). It follows that also part of this set of genetic attributes are those that act to maintain peace, under differing conditions. The ultimate goal of the entire set is survivial of the species. Manifestation of all genetic potentialities is shaped by past experiences (including learning), and present circumstances (Dittus 1975).

An animal's whole psychology and social structure is a secondary adaptation to its primary needs to feed and to reproduce, but 'on an average, social bonds will not interfere or run counter to the basic adaptation to the environment. The psychological and the social organization are the result of the adaptive syndrome to the environment' (Dittus 1975:236).

Shafton (1976:8) points out that natural selection endows an animal species with genetic means for developing an assortment of behavior systems which let it 'deal adaptively with a class of situation which recurs in the life of its species' in its particular niche, in either social or nonsocial environments. 'Taking behavior over time, it appears that the animal commences and terminates action only by separate cutting in and cutting out of a coordinated array of perceptual biases, motivational tendencies, and motor potentials' (Shafton 1976:8).

All of this knowledge raises questions such as:

> In what recurring circumstance in the evolutionary history of the chimpanzee, might negative, aggressive behavior – which destroys the society – be adaptive?
>
> What is the selective value in the ability of chimpanzees to switch very rapidly from the loosely structured mutual dependence system which is the normal adaptation and optimal strategy for foraging their natural habitat, to a rigidly structured, hierarchical form of social organization?
>
> What class of situation would recur that might render such social change and behavior adaptive toward survival of the species?

Of course, such questions cannot be answered adequately in the context of this study. They require deep and lengthy consideration. But the type of situation that is most likely to recur in the evolutionary history of the chimpanzee (and other forms of animal life on our planet) is probably a natural environmental catastrophe of some kind – flood, long-lasting drought and so on – which brings about severe food shortage. An individual, group or species cannot survive very long without adequate food. A very simple possibility is that under a very prolonged situation of extreme, unrelenting food crisis the open social system might become disadvantageous, and rapid social change to a different form of organization might, in crisis circumstances, be a survival mechanism.

Crises are of two sorts, natural (environmental) and cultural. It has been found that among humans, natural environmental crises, such as flood, earthquake, drought and so on, tend to unite people in the face of the common catastrophe. But a cultural crisis, 'the direct result of some dysfunction inherent in the very form and dynamics of a given mode of culture' involves the disintegration or suspension of some basic elements of the established way of life (Bidney 1967:348).

The (simulated) prolonged food crisis brought about by the methods in which the bait foods were supplied at Gombe instigated a change to a set of negative behaviors, which led to a cultural crisis. When the chimpanzees began competing directly and aggressively for resources, the old open, cooperative, nonacquisitive system no longer operated. *The structure of attention broke down.* The apes shifted their focus from monitoring what was happening in and around the group to their own concern to obtain some of the coveted, difficult to obtain, bait.

Chimpanzees are adapted to indirect competition for food, a separate, simultaneous seeking of the same food resources. The observers took control of access to a desired and scarce food and, through the frustrating and (to the apes) incomprehensible method of distribution, added the tension of anxiety over food to the artifical feeding situation. Unintentionally, the Gombe and Mahale apes were pushed into competing directly and aggressively against conspecifics for the bait foods. Perhaps the apes could tolerate this situation for a short period, natural environmental crises usually pass, or in the vanished uncrowded world, the animals had the option of migration. These groups do not have this option, and the frustration and anxiety over food was continuous over several years. In the sociopsychological sense, the Gombe chimpanzees experienced the anxiety of an unnaturally prolonged food crisis, together with an inability to use the adapted means of solution, fission, to resolve it.

When a situation with which a higher primate species is faced results in a radically different behavior system cutting in, a different mind set also is involved.[46] Perhaps this simulated dire food crisis triggered a switch to a negative mind set among the Gombe apes, in that the aggressive direct competition for bait-food spread to include almost all other aspects of the chimpanzees' social life. The males began to compete directly and aggressively with their former peers for monopolistic access to estrous females and top rank in a hierarchy based on dominance. At the group level, they competed for sole access to formerly open, shared territory. The behavior of the Gombe apes was negative, erratic and unpredictable. The benign, open, mutual dependence system no longer held. Some sort of social change was required, if the group was to survive. Here follows a tentative proposal.

It may be that suspension or de-emphasis of the positive set of behaviors imperative to the mutual dependence system, and change to a negative set which will act to break down the existing system, is a prerequisite to rapid change to another form of social system. But here we must digress, by way of explanation.

In their illuminating book *The Social Bond*, Nisbet and Perrin (1977) write of change, social processes and significant social change as three different processes. They indicate that by the term change they mean 'a succession of differences in time within a persisting identity', for example, the formation of a new family through marriage, parenthood etc. (Nisbet and Perrin 1977:226). This kind of process is not actual, fundamental social change because 'the basic structure remains intact' facilitating its persistence over time (Nisbet and Perrin 1977:267).

Nisbet and Perrin emphasize that, particularly in theories strongly characterized by biological premises, we find social change equated with growth or development. 'The key idea or assumption is that the cause of fundamental change inheres in social structure' (Nisbet and Perrin 1977:275). To the contrary, *substantive change always emerges as a result of crisis*, Nisbet and Perin (1977) indicate. (To these scientists, crisis is as W. J. Thomas defines it: a relationship between the individual and environment 'precipitated by the inability of the ... [individual] (or social group...) to continue any longer in some accustomed way of behavior' (Nisbet and Perrin 1977:281)). This sociological definition refers to crisis among human beings, but it is also useful in defining the crisis that the Gombe and Mahale chimpanzees experienced.

In effect, the prolonged stress of the restrictive feeding methods removed the Gombe and Mahale groups sociopsychologically from the relaxed (positive) social environment of indirect competition, and put them under abnormal stress for a very long period of time, which resulted in change to a negative mind set. A rapid, radical change took place in their mode of resource competition, from what Fretwell and Lucas (cited by Krebs and Davies 1981:94) refer to as 'ideal free conditions' in which all individuals are free to go wherever they like with equal access to resources, to the opposite model, 'despotism', i.e. exclusive use of territory through defense of resources. This is not the usual behavior of chimpanzees when food is short, which is common in the dry season. Normally, under such conditions, they forage separately and silently, and feed more eclectically, no longer going about in small subgroups, or food-calling.

It seems possible that quite unintentionally the Gombe feeding methods brought about the stressful emotive atmosphere of rare, acute food crisis such as might be brought about through either overpopulation or prolonged natural disaster, which may have made adaptive, under the special circumstances, a change to the mode of resource defense Krebs and Davies (1981) refer to as 'despotism': exclusion of others from resources. On the other hand, in the world in which their behavior evolved, under the stress of actual acute resource crisis, an obvious, simpler solution is for the group to split, long before such change is necessary, for some members to migrate to a less food sparse (less crowded) area, a simple group extension of the adapted pattern of individuals separating to forage (fissioning), in the dry season. Under the stress of natural environmental crisis, the local group may break up, but the basic social system does not change in form.

A very real problem with the above solution is that, because of the fission and fusion pattern and shifting, positive charismatic–dependent relationships, chimpanzees are highly attracted to the group as such, less so to specific others. Moreover, the very loose affinitive social bond to the group must also be exceedingly strong, persistently tenacious, if the fissioning and fusing group of autonomous individuals is to hold (see Power, 1988).

The question, then, is: what might be required (in a situation of lengthy food crisis) to induce a number of such pertinaciously bonded primates to break away from their group, to leave before they reach a state of malnutrition in which they are too weak to survive what might be an arduous journey? Violence, rejection and refusal of access to resources by others, formerly their supportive companions, might act as a catalyst. What this notion suggests is, when used within the group against each other, the selective value and function of aggressive behavior is to break strong emotive bonds to the group and so dissolve the strong relationships which form the adapted social structure. This clears the way for either spacing out, formation of a new group with the same social system, or a necessarily rapid, crisis-induced change to some form of organization which is advantageous in the survival-threatening conditions. *Aggressive behavior is functional when it brings about significant social change, when change is necessary, for survival.* By also closing their home range and its resources to outsiders, or by actually driving away (or exterminating) its weaker neighbors, and annexing their foraging range, a group strong enough to hold the extended territory and monopolize its resources has a better change to survive, when food shortages become long standing and acute, and the gains in resources exceed the costs of defense. In the vanished uncrowded world a group grown too large for the food resources could split, and some migrate to another area. Under natural conditions, it would not be optimal strategy to kill the dispossessed, as happened at Gombe and Mahale.

Which members might allow themselves to be driven out? It would not be adaptive if only the females and young and elderly left, without protectors and prospective mates. I suggest that some particularly independent charismatic members (of either sex) would prefer to leave, rather than take a position in a ranked hierarchy, and their leaving would attract followers of both sexes. They would move away in mixed groups, as complete, microcosmic, new societies. One might argue that in the normal social order, each small charismatic/dependent combination or subgroup of two to ten chimpanzees is a microcosm of the whole society. As newly formed local units, such split-off groups are able to migrate as far as necessary in order to alleviate whatever problem exists, while still retaining the adapted mutual dependence form of social structure.

The level of aggressive behavior demonstrated by the Gombe apes is far beyond that required to cause independent, self-reliant primates adapted to positive behaviors and a pattern of exiting (fissioning) from even mildly stressful situations to break away from their tense group. But, as we know, human-surrounded chimpanzees no longer have this option.

The chimpanzee hierarchy

Linear or near-linear forms of dominance hierarchy are a common phenomenon among many species of primates and other animals. According to Chase (1974), two models have been proposed to explain hierarchy formation. The first is a 'tournament' model, which suggests that animals win their places in a hierarchy through a 'round robin' competition against all other group members. These competitions may take the form of 'violent fights' for dominance rank, or of passive recognition of superior and subordinate rank (Chase 1974:374). The second explanation is a 'correlational model' (Chase

1974:378).[47] This model posits that there is a high correlation between any trait or any composite of traits assumed to predict dominance, and the actual position of animals in a hierarchy.

A dominance hierarchy among primates is usually assessed as a tournament model, in terms of alpha rank gained through aggressive direct competition. But dominance is better equated with *control* because, as Chase (1974) suggests, there are means other than aggressive competition through which control of a group may be gained.

Chase (1980) points out that while in the literature explanations of hierarchy are based on differences in individual characteristics among group members; in fact hierarchies emerge from the *interactions* of group members, not from differences among them. For example, when groups of unacquainted humans are assembled in either laboratory or natural settings, a differentiation of members quickly emerges 'on the basis of frequency of originating and receiving various kinds of behavioral acts and ratings of leadership and likability by fellow group members' (Chase 1980:908). Although one might expect such 'preference choices to yield structures considerably different from those produced by dominance relationships, there are considerable similarities' (Chase 1980:908). *More refined data might show that preference relationships and dominance relationships are identical*, Chase continues.

This leads me to suggest that there is normally an all but invisible status and role-based dominance hierarchy within chimpanzee local groups. Each charismatic–dependent subgroup is a self-contained epitome of this positive form of hierarchy. This positive form of social dominance order is based on preference. Authority is gained through attraction and voluntary support – cooperation. In normal circumstances the control element of the charismatic leaders is muted, appearing as willingly accepted, but otherwise unenforceable, influence.

Chimpanzees are normally independent, yet profoundly social, animals. The temporary leader of each of the many subgroups is attracted to a number of other (often older) individuals, and sometimes chooses to move about with one or another of these animals, thereby assuming the dependent role in a new charismatic–dependent alliance. Those now leading, also, are attracted to other (to them) more charismatic individuals. As suggested earlier, there are a few extraordinarily charismatic chimpanzees, usually confident, exemplary elders whom all of the members of the group like, respect and trust, and to whom all defer.

Chase (1980:923) suggests that 'interactions at a micro level – among individuals – can be cumulated to explain how a macro level structure – a dominance hierarchy – is produced.' Accordingly, it might be said that these few outstandingly charismatic members hold potential alpha rank in an informal, latent dominance hierarchy based on group preference, which exists as the practical outcome of a sum total of individual decisions. (Group response brought about by accumulation does not reduce the individual's freedom of choice).

Gombe's David Greybeard, and the elderly female Flo, Kortlandt's deferred to 'grand old man', Mama and Yeroen, the oldest members of the Arnhem zoo group, the wild females Pandora in the Guinea group studied by Albrecht and Dunnett, Gray, of whom Ghiglieri writes, and Kasonta, the much followed top-ranking male in the Mahale K group, come to mind as extraordinarily confident, charismatic chimpanzees.[48] The extraordinarily charismatic members are normally relaxed and nonaggressive, but under conditions of abnormally high stress in which the usual behavior patterns no longer function efficiently (e.g. crisis), it is adaptive (for reasons that will become apparent) that they are also capable of becoming assertive to the point of aggressive dominance. Writing on the psychobiology of aggression, Moyer (1976:3) points out that assertiveness (which he defines as being 'the positive affirmation of a point of view'), even if forceful, is not considered aggressive 'unless an intent to demean or otherwise hurt' another individual is involved. When the intent is to influence behavior constructively, threat or attack is purposeful rather than oppressive.

There is some evidence that some charismatics have such capability. Goodall (1971*a*) reports that the ape David Greybeard, who was normally so amiable, was also the chimpanzee she feared the most, as he could be a very dangerous animal. As we know, David was not easily roused to anger, but, when roused he was impressively determined, assertive and potentially, dangerously aggressive. (See also note 32).

De Waal writes of the enormous respect in which Yeroen, and to an even greater degree, Mama, were held by the group. Like Goodall, de Waal (1982) considers these two extraordinarily charismatic apes to be undoubtedly also the most dangerous animals in the zoo group. The elderly female Mama, when childless was 'the group leader, dominating not only the adult female, but the adult males as well' (de Waal 1982:56).

Disliked, feared apes – such as Gombe despots Mike and Goliath, who seized alpha dominance position through aggressive contests, and used their power oppressively to force submission on other group members – do not command the high positive respect and widespread *support* of their followers. They command widespread *submission*.[49] Kasonta is described by Nishida (1979) as an aggressive ape. We cannot tell to what extent his aggression was a result of crisis level stress brought about by abnormal tension over (bait) food, but 'earlier top-ranking males were far less aggressive' (Nishida 1979:94). Despite his aggression, Kasonta is the most followed, most mated, most groomed Mahale male, Nishida reports.

Although sociologists Nisbet and Perrin (1977) hold that while typically, informal (charismatic) authority is one of influence, it is also 'inseparable from the situations in which it becomes manifest.' Accordingly, it may in some situations 'reach despotic proportions' (Nisbet and Perrin 1977:112). 'As a general rule, authority tends toward formality (structure and regulation) as the practical necessity for coordinating activities and functions increases' (Nisbet and Perrin 1977:112). It follows that major (significant) changes in societies are 'simply incomprehensible' except in terms of the crisis role of 'superlatively endowed' individuals, 'working within social circumstances, usually those of crisis' (Nisbett and Perrin 1977:284).

Nisbet and Perrin's hypothesis refers to social change among humans. Nevertheless, in times of dire food crisis, if the ideal (open) free mode of foraging typical of undisturbed chimpanzees is no longer the best strategy for survival, extraordinarily charismatic chimpanzees have the personality qualities necessary to take on a more authoritative form of leadership: alpha in a linear hierarchy, dominating all others. Such outstanding, respected individuals have, and can be expected to retain, the support of the group.

In 1968, Gartlan suggested that a dominance hierarchy is a natural form of organization which might appear under significant stress. If the best means of survival involves closing the group and its territory, a structured form of organization is required, to enforce these conditions. In crisis more than at any other time, if a group is to survive the leader must command obedience, must dominate by one means or another. Only one or a few extraordinarily charismatic members – human or chimpanzee – have both the long-established respect and deference of the entire group, and the confidence and strength of will which will permit them to dominate; to be as authoritarian as necessary. When physical force is required, they are capable of that too, and they have the backing of the group to enforce their protective domination.[50]

The proposal put forward here is that when circumstances render it necessary, a change from a loose, almost invisible, correlational hierarchy based on preference, to a structured dominance rank order, can be effected without a change of alpha leader and without a 'tournament' based on aggressive competition. Such evolution under crisis conditions is not, in Nisbet and Perrin's terms, a substantive social change, but a rare, radical *social process*, in that, in terms of function, it has a social purpose, hence the basic mutual dependence structure would remain intact. I further suggest that the dominance hierarchy as seen at Gombe and Mahale is a distortion of a rare but not abnormal social process. Free-living chimpanzees whose habitats are surrounded and limited by human settlement no longer have the less drastic option of the group dividing and excess population migrating to a new home range, distant enough to relieve overpressure on the local food supply.

When roused as a response to actual acute crisis, aggressive behavior may be adaptive, as a means of dividing a very cohesive group. It is not normally adaptive once the crisis passes; hence when resource stress is relieved and tensions ease, we might expect the customary positive, pro-social behaviors to reappear, over time. After all, when not aroused to aggression, it is characteristic of most charismatic chimpanzees, particularly the extraordinarily charismatic members, to be relaxed, amiable and nonacquisitive. Moreover, under normal conditions a structured linear dominance hierarchy is not the optimal solution to socioecological problems of chimpanzees. It seems likely that when the crisis is finally over, in a habitat in which food is patchy the open, egalitarian, self-directed way of life will reappear over time, although this we can but muse on.

What is more impressive, and profoundly important to our understanding, is the behavioral flexibility regarding structure and adaptation that the Gombe and naturalistic field studies, taken together, demonstrate. The contrasting findings are strong indications that theories tracing the behavior of chimpanzees – and humans – directly to evolutionary and genetic causes are sometimes misleading.

Further suggestions and comments

The Rousseauean foragers: human and chimpanzee

In 1762, Jean-Jacques Rousseau (who was not a theorist of complex modern nation-states) envisioned an ideal egalitarian society in which no one person held authority over others, and all had equal rights. He theorized that the best means for humans, for survival of the species, was to develop a system through which people could come together, unite as a group and benefit from their pooled strengths. Although the clauses of the 'contract' need not necessarily be formally enunciated, they must be everywhere the same and everywhere tacitly admitted and recognized. What is needed, Rousseau argued, is for humans to find, and mutually agree on, a standard to govern their ways of living without giving up their natural liberty to any person other than themselves as a collectivity. The problem is to devise a social system which will defend and protect each member with the whole force of the group, while allowing each individual to retain the full autonomy that they have outside the group.

Rousseau's envisioned system, his 'social contract', cannot change natural physical inequalities (such as size, strength, intelligence and so on), but it substitutes an equality whereby all members become equal by the natural rules of the act of association. These conventions will be obligatory only as long as they are mutual, Rousseau asserts, and their nature is such that each individual acquires the same rights as are yielded to others. In fulfilling the customary agreement, both self and others must benefit equally. Further, if the social contract is violated, each associated individual at once resumes his or her natural liberty, by the mere fact of rejecting the agreed contract for which he or she had constrained it. To make Rousseau's system work, every member's goal and self-interest must be the same; and social practices must benefit all, be to everyone's mutual advantage. These are the principles on which the mutual dependence system depends.

Rousseau's 'social contract' can certainly be viewed as ethical. Chimpanzees are not moral, they do not have values in the same sense as do humans. But they know what is socially acceptable and not acceptable. They know when and how to control their impulses, and what is required of a member. They have a practical, cognitive appreciation of the cultural/social expectations. Their constraints are adaptive, fitting, and useful – practical, not moral constraints – sometimes proximally restraining the individual's interest, but ultimately, in pursuit of individual interests, goals and benefits.

Problems with Rousseau's ideal social system have been solved by the chimpanzees (and the human foragers) through the adapted mutual dependence system. In renouncing the 'contract' with a particular charismatic leader or dependent follower, the larger system is not disturbed. The social contract continues with some other, more compatible leader, or follower.

A goal or goals, defined as being one or more condition, purpose or object sought to satisfy a need or want (thus *motivating behavior*) is common to all individuals and social groups, animal and human. Indeed the sociological view is that the goal is the reason for existence of a social group, even if not the official or recognized purpose. Fulfillment of the needs for food and reproduction are primary, but at another level a goal may be the maintenance of a continuing state, such as peace, or self-esteem (Barkow 1980).

Through the principles of autonomy, egalitarianism and non-interference mutualism implicit to the charismatic–dependent relationship – all operating at both individual and group level – the mutual dependence system assures a state of continuing self-esteem. That such principles lead to a normal or healthy state of self-esteem is implied in the (earlier mentioned) remark of a !Kung informant to anthropologist Lee (1979:348) that every San adult is 'headman over [herself or] himself.'

Confidence and self-esteem are related attributes, characteristic of 'true' (charismatic) leaders. With few exceptions, all adult foragers and chimpanzees experience the leader role and status from time to time. Hence it seems reasonable to assume that chimpanzees experience a state of psychological well-being tantamount to self-esteem, which is at the same time an unconscious sociopsychological goal and reward of the mutual dependence system.

If this assumption is valid, then unlike goals and rewards in the complicated, complex societies, the goals and rewards of all members of a mutual dependence society are not only equal, but the same. Such goals do not come in a form that can be monopolized. Further, what benefits the individual benefits the group, and vice versa. Thus the mutual dependence system involves little reason for conflict or division.

I am aware that, oriented as we in the West are to power, direct competition, initiative and aggressive pursuit of 'success', many readers will have – as I had, initially – difficulty with the argument that the fundamental adapted form of social organization of humans and chimpanzees is egalitarian and based on positive behavior and a relationship of mutual dependence between autonomous actors shifting between fundamental leader–follower status/roles.

It is our perception of self-interest as normally selfish, i.e. concentrated on one's own advantage to the exclusion of regard for others –

which it cannot be if such loose social bonds between autonomous individuals are to hold – and of the struggle to survive as a contest for superiority that creates the difficulty with the idea of the co-function of the primary factors of dependence and autonomy and makes them appear to be incompatible. But self-interest can be enlightened (as in noninterference mutualism) and self-direction or autonomy can be as Rousseau envisioned (and undisturbed human foragers and chimpanzees demonstrate), an organizing and structural principle of the group.

The Darwinian struggle for existence is not by definition an aggressive or directly competitive struggle for survival. It is a struggle or competition for existence only in the sense that every organism strives to maintain itself. Montagu (1976) points out that Darwin himself said that, when he used the term struggle, he meant it in a large, metaphoric sense which includes dependence on another in a struggle against natural conditions unfavorable to the individual and the species.

The positive form of self-interest is not a struggle for superiority or advantage, but for personal well-being, self-esteem. If enlightened self-interest was normally a struggle for superiority – which involves higher and lower rank, privilege and status – the foraging system could not have endured, as it has from humankind's beginnings to a short 10 000 years ago. Nor could it endure among chimpanzees, as the disintegration of the Gombe group demonstrates.

Rousseau's theory has enormous appeal, but, in Western society it is generally viewed as overoptimistic, romantically ideal, unrealistic. However, the model of the mutual dependence system of foraging humans and chimpanzees is not romantic; even less so is it unrealistic, except in terms of Western values. The highly positive system of mutual dependence is simply the optimal strategy for survival of these two species of primate, human and chimpanzees.

Montagner and co-workers (1978, 1979) are interested in the ontogeny of nonverbal communication and the social behavior and organization of contemporary Western children (of 12 weeks to 6 years) in day care and kindergarten settings. Interestingly, and perhaps significantly, these scientists found among these children the same charismatic and dependent personality types interacting in the same fluid relationship between many charismatic (leader) children and attracted (dependent) followers, who move freely and constantly from one temporary group to another, that are postulated in this study among human foragers and also wild chimpanzees.

Montagner *et al.* found that leaders in the pre-school groups are children whom the researchers consider also as dominant on the basis of being attractive to others. They lead 'without there being necessarily an intention to lead' (Montager *et al.* 1979:26). Others imitate and voluntarily follow them. They are friendly, approachable and popular peacemakers, 'easily accepted by older children and adults in every social context' (Montagner *et al.* 1978:227). Like the young male ape Pepe, who so impressed MacKinnon (1978:76), these leader children too move easily through the various circles of the community.

Hold (1976) is interested in whether, and to what extent, the behavior of 'high-ranking' (popular, nonaggressive) human children is 'rank-specific'. She also chose as her subjects pre-school (3½–6 year old) children, on the assumption that their behavior is 'not yet as self-consciously controlled as that of adults' (Hold 1976:179). Hold determined that one behavior or function of the influential popular children is to act as protectors of the group. Although usually nonaggressive, these children are peacekeepers. They intervene in quarrels and threaten, scold, or hit children who attack others, as do charismatic chimpanzees. They comfort and reassure children distressed by an aggressive attack, usually by putting their hand caressingly on the child's back or shoulder. None of these positive behaviors is typical of the aggressive children, who are disliked and avoided.

That these separate studies report the same charismatic leader–dependent follower role behavior and group dynamics among young contemporary children, *whose cultures are not organized in the foraging manner*, strongly suggests that there is some genetic factor involved in such social interaction. These young children too seem to enact the generalized protective parent–dependent child charismatic leader – dependent follower roles.

Unlike pre-school age children, employees in large, modern business organizations cannot be said to be less socialized than, or not as self-consciously controlled as, (other) adults. Yet Wedgewood-Oppenheim, who has studied the structure and function of such organizations, indicates that an important characteristic of successful, 'excellent' firms is a set of values and a social climate 'that encourages independent action, personal autonomy, exploration and risk-taking' without reprehension 'if the risks do not pay off' (Wedgewood-Oppenheim 1988:313). He notes that such excellent firms parallel the chimpanzee and foragers' carnivals through frequent celebrations such as Friday afternoon 'beer busts,' sporting activities and so on, which bring the many employees together in an enjoyable, informal atmosphere. Wedgewood-Oppenheim (1988:319) suggests that this basic type of social organization found in successful businesses may not be a product of industrial society, as is usually assumed, but may be (as I suggest), 'based on fundamental social forces that human beings share with primates.'[51]

A mammalian system?

The mutual dependence system, which developed as a socioecological strategy for living by foraging, may be still broader than has so far been suggested. It may be a mammalian, rather than just a primate, system.

Eisenberg (cited by Saayman 1975) suggests that studies of 'large-brained, long-lived non-primate mammals', for example humpback dolphins studied offshore in the Indian Ocean, challenge the assumption that 'complex behavioral repertoires, communication systems, social traditions, and an advanced form of social organization are unique to the higher primates.'

Saayman's (1975) and Saayman and Taylor's (1973) studies report that large schools of these marine mammals fission into independent subgroups of one to ten, to forage and travel independently throughout a home range. From time to time the subgroups reunite as a large social group, at a good feeding ground. At these recurring larger group reunions, vigorous social interactions take place, which seem to operate as a form of greetings ceremony, functioning to reinforce social bonds, Saayman suggests. After a time the dolphins depart, 'in newly constructed subunits, which frequently alter in composition from one sighting to the next' (Saayman 1975:197).

Saayman (1975:197) recognizes that these observations of humpback dolphins show 'striking similarities' between their social organization and the flexible social systems characteristic of undisturbed chimpanzees, some social carnivores, and gathering–hunting 'Bushmen' as the San people were then labeled. Efforts to reconstruct the nature of the social organization of early hominids may be broadened by investigations of various animal species showing such organization, Saayman continues.[52]

Focus for the future: pro-social behavior

For too many years the focus of many of those interested in primate behavior and social organization, and the nature of human nature, has been on male behavior, aggression, rank and dominance, territoriality and direct competition for 'resources,' by which term is meant food and estrous females. Reproductive success cannot be measured in numerical terms of copulations or even impregnations achieved. For a primate individual or species to survive, the imperatives are cost-efficient access to sufficient food, and socioecological conditions that ensure the survival of offspring to breeding age: a healthy next generation.

The Gombe and Mahale studies are of immeasurable value toward our understanding of both cause and effect of antisocial and pro-social behavior in the higher primate species closest to humankind. Now we must turn our concentrated attention to positive or 'pro-social' behavior, 'helpfulness, generosity, kindness, cooperation, or other behaviors that benefit people [and chimpanzees]' (Staub 1984:ix). Such study has begun in several countries, but is in its early stages, Staub testifies. As Montagu (1970:29) emphasized, 20 years ago, 'it is in the methods of peaceful association that strength is accumulated, and it is in those of competition, struggle and combat that it is dissipated and wasted. Getting along with one's fellows has great adaptive value.' There is a great deal to learn regarding how positive, pro-social behavior comes about, how it is maintained, what prohibits it, its nature, motivation and goals.[53] (And, one might add, because of the high value placed on power, advantage, dominance and aggressive direct competition by advanced technological nations, there is little time left in which to do so). The study of pro-social behavior is an area in which anthropologists and primatologists can make important contributions, urgently required, in that we humans are a self-endangered species.

NOTES

1 The captive apes which de Waal studies in the Arnhem zoo live in excellent zoo conditions. They have far more space and more natural-appearing surroundings than do most zoo collections, but even in these superior conditions the captive apes cannot follow the natural pattern of fission and fusion. 'They can never completely isolate themselves from the group' (de Waal 1982:25). In other words, they cannot use the adapted method of fission by choosing to exit from the group and the vicinity for days or months at a time, which averts and disperses tensions between individuals before they build up to an explosive level and threaten the harmony of the group. Their inability to use this simple means of dispersing aggression at an early stage has had serious detrimental effects on the captive group.

I think now, with publication imminent, that more weight than is assigned it in this volume should be given to the impact on free chimpanzees in most places they have been studied (i.e. the Gombe and Mahale National Parks, and the Kibale Forest) of being now entirely surrounded, inhabitants of shrunken islands of wilderness in a sea of humanity. This has resulted in attenuation of the natural solution of fission, of spacing out to sufficient distance to avoid competition for insufficient food resources.

Moreover, I now realize that use of a broader situational term such as 'exigent' would have been more useful than the categorical term 'provisioned', in that it is both more apt and more inclusive. A category of exigency would cover all chimpanzee groups experiencing *abnormal* crisis (as distinguished from rare but *normal* geo-ecological crises such as flood, fire, drought and food crop failure). An *abnormal* crisis is defined herein as being a prolonged relationship between individual and social groups and environment, for which the species involved has either no adaptive response, or in which the adapted survival mechanism is unable to operate.

A chimpanzee group in normal crisis usually experiences food shortage, a problem which in the undisturbed, wide, wild world could be solved through implementation of various forms of fission. Voluntary or forced group [and individual] fission, and migration to a new location, relieve overpressure on an insufficient food supply and establish a new, independent yet familiar group, not in direct competition for food. (Forced migration is taken up in Part 7).

2 See, for instance, *Webster's Seventh New Collegiate Dictionary*, 1963; 222. *The Shorter Oxford Dictionary* (1958), p. 483, defines a dependent as 'a person who depends on another for support, position, etc,; a retainer, subordinate, servant,' and 'the condition of a dependent; subjection, subordination.' According to Bowlby (1973), the mentally healthy individual shows a mixture of self-reliance in some situations, and normal dependence in others.

3 The unanimity of the separate findings from the naturalistic chimpanzee field studies, of which Ghiglieri's is one, should have gained them a certain acceptance; however, the findings raise anomalies, i.e. violations of the paradigm-induced expectations (Kuhn 1970), which do not fit within the established dominance model. This is probably one, perhaps unconscious, reason why they have been downgraded and ignored. The rejection fits Kuhn's (1970) theoretical conceptualization of the resistance to paradigm challenge or change. According to Kuhn (1970:64), 'in the development of any science the first received paradigm is usually felt to account, quite successfully' for most of the behavior observed by that science's practitioners. Accordingly, the concept is refined and 'professionalized,' with the result that the scientists' vision is increasingly restricted and committed to the established model, which leads to 'considerable resistance' to paradigm challenge or change (Kuhn 1970:64).

4 Publication dates tend to follow long after the dates of actual study. For example, Goodall comments that her chapter 'Chimpanzees of the Gombe Stream Reserve' (Goodall 1965*a*) is based on data collected during several periods between June 1960 and December 1962 (Goodall 1965*a*:425). 'Mother–offspring relations in free-ranging chimpanzees' is based on data collected between 1960 and 1964 (Goodall 1967*a*:287). 'A preliminary report on expressive movements and communication in the Gombe Stream chimpanzees', (Goodall 1968*b*), is based on data collected between 1960 and 1965 (Goodall 1968*b*:314).

5 There are very few adult females who are both adult and without dependent offspring for any length of time. The brevity of the childless period severely limits their numbers in the wide-ranging swift-moving subgroups. Sterile adult females seem to be very few in number. Of 27 mature females (10 to 35 years of age) at Gombe, only one female, Gigi, was permanently sterile, although she cycled regularly (Teleki *et al.* 1976). Adolescent females are not immediately fertile after their first estrous period. Goodall (1965*a*) reports that there is a time lag of from 4 months to 2 years before they conceive.

6 See, for example, Theodorson and Theodorson (1969).

7 In writing of the relations between the 'split-off' Kahama group and their wild unhabituated neighbors further to the south (dubbed the Kalande

group), Goodall (1986*b*:228) refers to an 'invasion' of the split-off group's territory by this 'powerful' group. (Powerful in that the wild group contained ten males to the Gombe group's six). Yet, from her description, the intent of the 'invasion' may not have originally been sinister. It is entirely possible that the wild apes first came looking for friendly social contact, i.e. carnival. But the disturbed Kahama (split-off) group met them with apprehension and hostility. The first encounter observed between the nervous Kahama males and a group of the incoming wild apes was in November 1974. When the first 'stranger' appeared, the Kahama males responded by screaming, fleeing, displaying, 'uttering deep, fierce-sounding roar pant-hoots' and by throwing 'at least 13' large rocks downhill to where the strangers were calling and charging through the bushes (Goodall 1986*b*:488). On catching sight of the observers, the unhabituated wild apes retreated.

Also in 1974, the provisioned Kasekela (camp area) group were seen to respond to their wild neighbors to the north in exactly the same way. Later encounters between both provisioned apes and their wild neighbors are reported to have been increasingly aggressive in nature. This strengthens the current view that 'interactions between males of neighboring communities are typically hostile' (Goodall 1986*b*:493). (See also note 39).

8 Among chimpanzees, a typical act of copulation lasts about 6 to 15 seconds, a matter of coupling and a few thrusts by the male.

9 Feeding the chimpanzees within the human's camp, which surely was to the then little-habituated apes a totally strange, alien environment, may have changed their sexual behavior. For example, according to Goodall (1971*a*) the elderly female Flo, always the most sought-after sexual partner, was an almost daily visitor to Goodall's camp. In July 1963, when Goodall still fed the few chimpanzees who dared to enter the camp area by heaping bananas on the ground, Flo came into estrus. She came into camp followed by 'as many as fourteen males,' many of whom, for the first time, 'braved' the novelty of the camp in order to keep close to Flo (Goodall 1986b:446). Goodall did not see any aggression towards Flo, yet on the last day of an unusually long (5 week) estrous period, the old female arrived in camp 'incredibly tattered,' cut, bruised and exhausted (Goodall 1971*a*:93–4). Goodall comments that she does not know why the males were so rough with Flo. As all males mated with Flo whenever there was any social excitement, over a span of 5 weeks the sheer number of matings would surely be exhausting. Goodall (1986*b*:446) reports that when Flo again cycled in 1967, she was observed to copulate 50 times in 1 day. It seems only reasonable to assume that this was not normal sexual behavior, but probably a result of social excitement, perhaps in 1963 through tension due to the human presence (in 1967 over the intense direct competition for the withheld bait food). It also suggests that the social behavior of the Gombe apes was being affected by human interference at a much earlier date than my chosen 'watershed' date (1965).

10 Field study of the free chimpanzees ranging the Kanyawara section of

the Kibale Forest was begun in 1983 by Dr Isabiriya Basuta. Following Basuta's 2 years of study, Dr Richard Wrangham carried out 2¼ years of study of the same group. These studies reveal that, although not artificially fed, in their inter-individual relations the Kanyawara (Kibale) apes are 'extremely similar' to the artificially fed Gombe and Mahale chimpanzees. Wrangham reports: (1) obvious tension between males when they meet, often leading to outbursts of aggression; (2) that most adult males compete repeatedly for status in a 'marked' dominance hierarchy; (3) inter-individual relations among the males of the group Ghiglieri studied for 18 months between 1977 and 1978, 8 kilometers away in the Ngogo section of the same forest, are *essentially the same* as those of the Kanyawara group (Wrangham, personal communication, January 1990).

This does make it appear, in that the Kibale apes have not been artificially fed, and the behavior of all four widely different groups is strongly similar, that such behavior is the adapted behavior of chimpanzees in general. However, in his doctoral thesis, Ghiglieri (1979) reports the Ngogo chimpanzees as being relaxed, gregarious, nonaggressive animals, living in open groups, without any indication of a dominance hierarchy. This being so, then there has been significant negative social change in the behavior and organization of this particular Kibale group since the late 1970s.

There is no way of knowing if there has been similar negative change in the behavior of the Kanyawara apes, but both groups face the same human-imposed problems, which have radically changed their environment (and may explain the negative behavior and organizational change in the Ngogo group, in a little more than a decade).

Ngogo is in the heart of Kibale Nature Reserve, hence the Ngogo apes were little disturbed when Ghiglieri studied them in 1977–8. Even at that date, the habitat of the Kanyawara chimpanzee ranging north of the Nature Reserve had been 'severely modified' through a government-sponsored program of logging, burning and selective poisoning of some 'weed' tree species, some of which were sources of chimpanzee foods (Ghiglieri 1988). As this is an ongoing (70 year) project, it has undoubtedly penetrated the forest further, wherever it is being carried out.

Uganda's human population doubled to 14 million between 1960 and 1980. One result is human settlement spreading on to the Kibale grasslands surrounding the forest. By the end of the 1970s, there were at least 7000 people living in the area, 1800 within the forest itself, largely in the south (Ghiglieri 1988:81). The Nature Reserve, which contains about 23 square miles (5960 ha), is judged to be large enough to support only two communities of about 55 chimpanzees each (Ghiglieri 1988:81). The ratio of monkeys to chimpanzees in the Reserve is 197:1 (Ghiglieri 1988:33), thus there is serious competition for food. As they are surrounded by human settlement, and logging is reducing the forest area, the Kibale chimpanzee population is being concentrated into an ever-diminishing area, in increasing direct competition for an (also curtailed) food supply. Obtaining sufficient food has become a problem, and the chimpanzees can no longer take the once available natural solution of the group fissioning, and some moving away to a new home range, distant enough to alleviate the direct competition for an insufficient food supply.

The four studied groups – at Gombe and Mahale and, by the mid-1980s, Kanyawara and Ngogo – share two common elements. One is the prolonged unnatural stress over an uncertain food supply (although at Gombe and Mahale this was largely through the wholly abnormal situation of the humans taking control of access to present and desired bait foods away from the chimpanzees). The other shared element is the already mentioned inability of the competing chimpanzee groups to move far enough apart to end the tension of direct competition, to which they are not adapted.

Being without this tension and competition-reducing solution, one might anticipate even more aggressive intra- and inter-group relations among the Kibale apes, in future years; possibly including some forms now reported among the Gombe and Mahale chimpanzees – murder, infanticide and cannibalism. Only time can test this supposition.

11 The question remains: did Goodall initially misjudge the age of the infants born before her arrival, or, in the tense and threatening Gombe atmosphere, were they slower in maturing socially than is normal?

12 Dr Shirley Strum found that, in order to save a baboon troop she was studying from extinction by incoming farmers, it was necessary to move the troop far from the vicinity of the farms. She had to catch and cage the wild baboons temporarily, for transport. One result was that although 'nursing and weaning normally flows smoothly and easily, following their own rules,' the caged, abnormally stressed mothers began attacking their babies, 'weaning became violent' (Strum 1987:234). As a result, 'the babies became even more insecure and sought further contact (with their mothers), beginning a vicious circle' (Strum 1987). Relaxed mother–infant nursing and contact was restored, when the troop was released in their new home. 'The severe and aggressive weaning . . . was gone, as were the frustration and signs of depression' (Strum 1987:240).

13 Interestingly, Turnbull (1965:306) reports that by 6 years of age both Mbuti boys and girls begin to go off on hunting and gathering forays on their own, in the vicinity of the camp. Usually, by the age of 9 years they accompany the adult hunting and gathering groups, and increasingly assume, or are permitted, adult responsibilities and privileges.

14 The migratory category of the Gombe census records, 1963–73, includes 'all unaccountable disappearances and all new chimpanzees who were previously unknown to human observers' (Teleki *et al.* 1976:562). Teleki *et al.* (1976:562) point out that both categories (immigration and emigration) are somewhat flexible, because some immigrants 'may have been familiar to named (habituated) chimpanzees.' I see no problem with this. Immigrants are not by definition strangers, but simply those who

migrate to another country (or habitat) which is not their own, for the purpose of permanent residence. In the case of humans, people tend usually to migrate to places where they do have acquaintances or relatives, and so are not entirely strangers. Perhaps chimpanzees also do so.

15 I thank Dr M. R. A. Chance for suggesting the very fitting term apprentices for the youthful wild chimpanzees.

16 In his various writings Ghiglieri (1979, 1984, 1988), indicates that, because he never followed the Kibale chimpanzees to the borders of their habitual range, he collected no direct observational evidence of their territorial behavior or interactions with neighboring local groups. He also makes clear that he saw no real aggression between the wild apes at any time during his 18 month field study.

Yet, in his 1988 book, he states that the Kibale males 'consistently behaved with such solidarity toward one another as to appear *prepared* for war at any time' (Ghiglieri 1988:258). 'Not having seen active territoriality does leave the question in Kibale open,' Ghiglieri (1988:257) admits, 'but no other explanation accounts for the striking differences I observed in activity patterns between males and females.' Another explanation is possible and offered in Part 5 of this book: parental roles. Ghiglieri (1988:258) continues, 'So, in conclusion on the question of whether territoriality is the standard pattern of wild chimpanzees, unprovisioned chimpanzees, the data from Kibale suggests it is.'

One must comment upon the unscientific irresponsibility of this assumption which, by his own report of not having seen active territorialism (see above), is not based on his Kibale data, or undisturbed chimpanzees (see note (3)). Ghiglieri (1988:260) then suggests that we humans 'quite likely share an unusual [genetic] legacy of cooperative male aggression against alien males.' In response to Ghiglieri's rhetorical question 'Does war run in our genes . . .?' the answer is, flatly, no. What we humans actually share with the chimpanzees is a *potential* to be aggressive (and territorial) or peaceful (and open), depending upon the circumstances. (See also note 41).

17 I suggest this because anthropologists find that human local foraging groups are units in a larger interacting, interbreeding population. Recently, I learned that Sade (1972) suggested that the basic unit of nonhuman primates is not the single local group, but a system of local social groups, interacting in a given area.

18 Goodall (1986*b*:162) reports that large, very social 'gatherings' *limited to members of the habituated Gombe group*, still take place (my emphasis). (See also Part 6, The social function of the carnival reunions).

19 They seem also to have some aversion toward unrelated fellow group members. In her later book, Goodall (1986*b*:176) does indicate that while male siblings typically support each other during social conflicts, the relationship between two unrelated male members of the group 'is likely to run a rather different course.' Youthful Gombe males usually challenge unrelated older males for dominance rank with much

aggression. Their relationship may become very unfriendly and stay that way for 'years rather than months' (Goodall 1986*b*:177).

The male Everad was forced out of the camp area group for many months, by aggression from allied brothers Figan and Faben. Everad was seen several times in the home range of the wild Mitumba group, apparently looking calm and relaxed. On at least one occasion he was accompanied by a wild female in full estrus (Goodall 1990:85–6). If chimpanzees in general share an aversion to strangers, one might have expected Everad to be tense and nervous in such circumstances.

20 The definition of the term role is adapted from Theodorson and Theodorson (1969).

21 The word 'policy' is used by de Waal (1982) to denote 'a consistent social behavior with a view to achieving a certain aim,' whether this behavior is determined by innate tendencies (intuitive policy) or by experience and insight (rational policy), or both.

22 The oldest member of the Arnhem zoo group, the female Mama, at 40 or more years, was clearly an extraordinarily charismatic animal. De Waal (1982:54–6) describes the slow-moving old female, as having 'great power' (confidence?) in her gaze, being enormously respected, and having a central role in the group through her 'stabilizing and reconciliatory influence.' When tensions in the zoo group build up to the point of physical violence, the combatants, often both, run screaming to Mama, 'even the adult males' (de Waal 1982:56).

23 'Following' is not physical in the sense of the leader being first in line of a moving subgroup. The term is used in the sense of some ape or apes choosing to move about with another for a time, endorsing this animal's leadership by going where he or she goes.

24 Even the very confident David Greybeard and his nervous-aggressive dependent companion Goliath changed roles when a large male baboon, competing for bananas, attacked David. David ran and hid behind Goliath who took the role of protector, attacking the baboon (see Goodall 1971*a*:82).

25 There are elements of both parental roles in the charismatic–dependent relationship. This hypothesis of course requires a great deal of further careful testing, as do all hypotheses offered in this study.

26 Any parent who has taken their very young walking infant for a stroll will recognize this very common, non-pressuring control technique.

27 The duration of human foragers' sanctions is dependent on the sanctioned individual's response. If a negatively sanctioned (shunned, ridiculed or ostracized) individual discontinues his or her disapproved-of behavior, the sanction ends. As long as desirable or approved behavior continues, the actor continues to receive a positive emotive reward through the approval, respect and acceptance of the group or society.

28 Although the zoo situation is not normal for chimpanzees, it is interesting that records kept at the Arnhem zoo show the 'low mating frequency of Luit in the months before his death' (at the hands and teeth of other males of the group). (See Figure 3 of de Waal 1986:247).

Unfortunately, de Waal (1982) does not indicate whether this was due to Luit's sexual interest dropping, female refusal, or his being excluded from mating by the males. (De Waal (1982) reports that Luit had been very active, sexually, in earlier years).

29 There is no evidence for this sort of group sanction in the studies of wild chimpanzees. However, the Gombe female Passion, who had cannibalized four infants born in the disturbed group, disappeared from the area for some time in 1980 and returned with 'extremely severe injuries' that appeared to be the work of chimpanzees (Goodall 1983*b*:34). Whether or not she killed, or attempted to kill an infant while visiting a wild group and was attacked by members of that group, will never be known. But, it is reasonable to assume that, as among human foragers, such harmful deviance would normally result in extreme sanctions.

30 Moreover, the emotional rewards appear to be two directional. When two apes meet and embrace, or one begins to follow another, it is not easily discernible which is initiating or expressing or positively sanctioning the behavior of the other. When one ape begins to follow another, is the follower expressing a desire for the companionship of the chosen leader, or is the leader positively sanctioning the actions of the follower by permitting him or her to be a companion? Detecting positive sanctioning is further complicated in that both actors in this NIM type of interaction receive pleasurable emotive reward.

31 Dr Chance's studies were largely of captive primates. Field studies of wild baboons by Dr Shirley Strum contradict the past picture of baboon society being based on a rigid, aggression-based dominance hierarchy headed by a dominant male. Strum (1987:144) found that a wild troop is organized around a complicated, sophisticated structure based on 'social reciprocity,' which necessitates the animals being 'nice' to one another. Real power resides with the 'wise,' not the strong baboon (Strum 1987:151). Nevertheless, the principle of avoidance of the feared, and attraction to the respected/liked, holds.

32 Interestingly, Ghiglieri (1988:262) writes of one male's 'exaggerated aloofness over the petty affairs of lesser chimpanzees.' It may be recalled that it was the charismatic Gombe male David Greybeard who led the first of a series of determined group attacks against the despot Mike in 1964 (Goodall 1986*b*:61). We have no way of knowing whether David was consciously enacting the peace-keeping role of a true leader, responding to the increased tension in the group caused by Mike's behavior, or whether the normally calm and gentle male was himself directly roused to anger by Mike's aggressive noisy displays and his unprovoked attacks on females and young, and responding as a disturbed individual – who was spontaneously joined by other individuals who also were distressed and angered by Mike's actions. As either a group response, or the sum of several individual responses, the function of the joint attacks was the same: to deal with Mike's society-disturbing behavior.

33 The only free-ranging chimpanzee group in which females are known to live alone with their offspring, quite separate from the males and other

females, is the latter-day Gombe group (cf. Wrangham 1979). Wild females are not reported to live separately from the males (Ghiglieri 1979), nor do the provisioned Mahale females (Nishida 1979).

34 In de Waal's (1982) opinion, this is an entirely conscious peacekeeping role behavior. (Peacekeeping is not specifically a female role, it is role incumbent upon all adult foragers, both human and chimpanzee). Perhaps only the Arnhem females carry it out because they are not involved in the male dominance struggle that goes on in the stressed zoo group. (If an Arnhem male did this, his action might be misconstrued as allying with one or other antagonist).

35 At the time of which de Waal writes, there were 23 apes in the zoo colony. Only four were adult males. The remaining 19 animals were adult females and infant or juvenile offspring. Therefore, distinguishing Mama as the leader of the females is somewhat misleading. She was leading the group, which happened to be largely female. Through the group endorsing Mama's action by backing it, social control was maintained. Without the support of the group, Mama's action would have been ineffective.

36 Kortlandt did not observe the high amount of social interaction among relaxed wild apes that is reported among the males of the Gombe and Mahale groups. Rather, he reports that individuals and small groups usually pass each other 'like commuters at a railway station,' each intent on his or her own pursuits (Kortlandt 1962:132). It is possible, even likely, that the competitive provisioned males must continually demonstrate their goodwill toward their uneasy fellows. Within a primate group which is relaxed, and the members of which are not competing directly against each other, there is not this need for continual reaffirmation of affinity.

37 Nishida and Hiraiwa-Hasegawa (1986:176) write of the one-way grooming relationship between the 'alpha' (aggressively dominant) male and Kasonta, the 'top-ranking' male, who had influence through his power to attract followers and mates, a 'nonintentional' (clearly charismatic) leader (Nishida 1979:93, 94). Despite the dominant (alpha) frequently grooming this charismatic male, Kasonta rarely returned the alpha's grooming (Nishida and Hiraiwa-Hasegawa 1986).

38 After carrying out separate studies of the Gombe apes between 1968 and 1975, Tutin and McGinnis (1981) conclude (tentatively, because their samples were small) that female chimpanzees generally prefer to mate with males who show high frequencies of affiliative behavior such as grooming and food-sharing. Such males gain increased reproductive success at the expense of males who show lower frequencies of affiliative behavior, Tutin and McGinnis suggest. In contradiction to the old saw, in the important area of reproductive success, 'nice guys finish first!' Tutin and McGinnis also report that adult males show a definite preference for mating with older females, rather than the younger ones.

39 Goodall (1986*b*:503–6) documents chronologically the steady deterioration in social relations between local Gombe groups, from the

summer of 1960, when she watched as 'excited and noisy' groups from the north and the south charged into a valley where figs were abundant, met, mingled and fed together peacefully to the increasingly hostile encounters of the 1970s, which ended in the complete annihilation of the Kahama (southern) group by the northern (camp area) group in late 1977.

40 Incest, or attempts at incestuous couplings, were not seen among the Gombe apes in the early years. But by the 1970s (infrequent) cases of incest were observed, sons with mothers, siblings, fathers with (known) daughters. In all cases in the first two categories the females tried to resist (Goodall 1986*b*).

41 For instance, the changed attitude of the frustrated Gombe apes toward outsiders, from social carnival to lethal primitive warfare, might be understood in terms of the theory of cognitive dissonance (see Festinger 1957, 1962). Festinger suggests that individuals are most at ease when there is internal harmony or consistency between their cognitions (feelings, attitudes, ideas, beliefs and so on), or a cognition and their behavior. When two simultaneously held cognitions are incongruous, this produces a psychological state of dissonance, which results in tension. Festinger theorizes that when people are induced to behave in a manner which does not fit with their cognitions, they will change their attitude, to bring the belief or feeling in line with the implications of their behavior.

Chimpanzees are a highly social species. The Gombe apes are attracted to others of their species, but because they have moved to direct, aggressive competition, which induces distrust, they are at once attracted to, and distrustful and fearful of, others. Hence they seek out outsiders, neighbors, but, in a reversal of attitude, anticipate and make war, not carnival.

42 It is zoologist Thelma Rowell's impression that the death of elderly females precedes or precipitates a change in behavior in blue monkey groups (T. Rowell, personal communication, 1989).

43 De Waal (1982:37) reports that sometimes aggressively displaying Arnhem males will pick up a stone or a stick and throw it at others. However, frequently an adult female is seen to walk up to the male and 'calmly' remove the object from his hand. De Waal does not identify which females do this, but it is reasonable to argue that such an action reflects self-confidence, rather than special authority (see note 34).

44 The ape McTavish is judged to be a charismatic leader on the basis of the following evidence:

1. He was an extremely unaggressive ape, second to the bottom on Albrecht and Dunnett's (1971:35) scale of aggression.
2. He was calm and unperturbed when the observers placed a dummy leopard in the observation field. He did not participate in the attack on the leopard as did some other apes, yet he did not retreat, and was

observed eating grapefruit nearby, only 5 minutes after the dummy was exposed (Albrecht and Dunnett 1971:109).

3. He was an elderly animal with the confidence to discipline a very confident adult leader (Pandora), which a nervous ape would be unlikely to have. When the deferring chimpanzee and the older ape he or she defers to are both of adult rank, the interaction is an expression of status, not of rank, difference, as it is also, among human foragers. Pandora and McTavish both have the qualities of charismatic leaders, but McTavish also has elder status and authority.

45 Both gave live birth when in their early forties. Graham's (1979, 1981) findings, cited by Goodall (1986b:86) show that in captivity some females continued to cycle regularly up to 48 years of age, 'although they conceived less often than when they were younger.'

46 Chance (1984, 1988) recognizes that two very different modes of mental operations (or mind sets) are possible among higher primates, with the result that one of two attached, antithetical types of social systems, agonic and hedonic, develops. An agonic social system is hierarchical, based on dominance through aggressive means. Individuals are preoccupied with potential threats to rank or their physical selves. Because their mental energies are focused on self-security, the free operation of their intelligence is restrained (Chance 1988).

On the other hand, a hedonic social system is based on positive relationships of mutual support (mutual dependence). Individuals are free from the fear of attack from others (and the struggle for rank position), so are able to give free rein to their intelligence, including their system-forming capabilities (Chance 1988).

47 The reader hardly need be reminded that the term correlational implies a reciprocal or mutually dependent relationship.

48 An impression all observers record is the strong personalities of these particular chimpanzees.

49 Indeed, as both aged, they slipped rapidly to the bottom of the distorted hierarchy, submissive to all of the stronger members. Goliath left the camp area (Kasakela) group and joined the split-off (Kahama) group. In 1975 a group of five Kasakela males patrolling encountered Goliath and beat, bit and pounded the 'extremely old' appearing former despot for 18 minutes, inflicting numerous severe wounds (Goodall 1986*b*:509). Although later the Gombe observers searched intensively for him, Goliath was not seen again, and is presumed to have died of his wounds.

50 Charismatic female chimpanzees have the same personality traits for authoritarian dominance as have charismatic males. The heavy predominance of childless adults available to take the active dominant leader role being male gives a misleading impression that this is a sex-based (male) role. Few adult female chimpanzees are available, i.e. are childless for any length of time, to take on the conspicuous active leader role.

51 Large business organizations require a formal hierarchical structure of positions; however, Wedgewood-Oppenheim (1988:313) suggests that in 'excellent' successful large organizations, 'the formal hierarchy is played down in favor of a very rich pattern of communication,' and that a high value is placed on individual contributions wherever in the formal hierarchy the contributors are positioned. Accordingly, there is 'constant regrouping' when new problems and issues appear and different skills are required (Wedgewood-Oppenheim 1988:315). Recent findings regarding the pesonality attributes common to the chairperson type of leader of successful management teams in business firms are also intriguing, as Chance (personal communication) points out. Typically, the most successful management teams are those having as leader a calm, patient, tolerant but commanding figure, who trusts and respects others and who generates their trust and respect in return (Belbin, 1989). These are, of course, attributes also of charismatic chimpanzee leaders, although they tend to be confident, rather than overtly commanding. I have suggested that they have the capability to become assertive, and commanding, if the necessity arises.

52 Russell and Russell (1979) cite Saayman's study in 1971 of baboons living in a natural reserve in the Transvaal. Saayman found that the least aggressive of three males in the troop was 'top leader.' This animal was the most sought out (and mated) by estrous females, and he groomed and was groomed more than the other males. He meets the criteria for being a charismatic leader. The most aggressive male 'scored low on all (of these) counts,' Russell and Russell (1979:110) report.

53 De Waal (1989) recognizes this need.

REFERENCES

Albrecht, H. and Dunnett, S. C. (1971). *Chimpanzees in Western Africa.* Munich, Piper.

Anderson, A. A. (1968). Discussions. Part III: Analysis of group composition. In *Man the Hunter*, ed. R. B. Lee and I. DeVore, pp. 150–5. Chicago, Aldine.

Averill, J. (1973). Personal control over aversive stimuli and its relationship to stress *Psychological Bulletin* **80**:186–303.

Azuma, S. and Toyoshima A. (1962). Progress report of the survey of chimpanzees in their natural habitat, Kabogo Point area, Tanganyika. *Primates* **3**:61–70.

Baldwin, J. D. and Baldwin J. I. (1981). *Beyond Sociobiology.* New York, Elsevier.

Baldwin, L. and Teleki, G. (1973). Field research on chimpanzees and gorillas: an historical, geographical and bibliographical listing. *Primates* 14(2–3):315–30.

Baldwin, P. J. (1979). The natural history of the Chimpanzee (Pan troglodytes versus), at Mt. Assirk, Senegal. Ph.D. thesis, University of Stirling, Scotland.

Barker, R. G., Dembo, T. and Lewin, K. (1941). Frustration and regression: an experiment with young children. *University of Iowa Studies in Child Welfare* **18**(1): xv, 314.

(1943). Frustration and regression. In *Child Behavior and Development*, ed. R. G. Barker, J. Kounin and H. F. Wright, pp. 441–83. New York, McGraw-Hill.

Barkow, J. H. (1980). Prestige and self-esteem: A biosocial interpretation. In *Dominance Relations*, ed. D. Omark, F. F. Strayer and D. G. Freedman, pp. 319–32. New York, Garland.

Barner-Barry, C. (1986). Rob: children's tacit use of peer ostracism to control aggressive behavior. *Ethology and Sociobiology* 7:281–93.

Belbin, R. M. (1981). *Management Teams.* Oxford, Heinemann.

Berelson, B. and Steiner, G. A. (1964). *Human Behavior*. New York, Harcourt Brace and World.

Bernstein, I. S. (1964). Role of the dominant male rhesus monkey in response to external challenges to the group. *Journal of Comparative Psychology* **57**:404–60.
(1966). Analysis of a key role in a capuchin (*Cebus albifrons*) group. *Tulane Studies in Zoology* **13**:49–54.
Bidney, D. (1967). *Theoretical Anthropology*, 2nd edn. New York. Schocken.
Bowlby, J. (1969). *Attachment and Loss*, vol. 1. London, The Hogarth Press.
(1973). Self-reliance and some conditions that promote it. In *Support, Innovation and Autonomy*, ed. R. Gosling, pp. 23–48. London, Tavistock.
Broom, D. (1981). *Biology of Behaviour*. Cambridge, Cambridge University Press.
Brownlee, F. (1943). The social organization of the !Kung (!Un) bushmen of the north western Kalahari. *Africa* **14**:124–9.
Buirski, P., Plutchik, R. and Kellerman, H. (1978). Sex differences, dominance and personality in the chimpanzee. *Animal Behaviour* **26**:119–23.
Buss, A. H. (1973). *Psychology: Man in Perspective*. New York, Wiley.
Bygott, J. D. (1972). Cannibalism among wild chimpanzees. *Nature* **238**:410–11.
(1979). Agonistic behavior, dominance and social structure in wild chimpanzees of the Gombe National Park. In *The Great Apes*, ed. D. A. Hamburg and E. R. McCown, pp. 405–27. Menlo Park, CA, Benjamin/Cummings.
Cancian, F. (1968). Varieties of functional analysis. *International Encyclopedia of the Social Sciences* **6**:29–43.
Carpenter, C. R. (1965). The howlers of Barro Colorado Island. In *Primate Behavior*, ed. I. DeVore, pp. 250–91. New York, Holt, Rinehart & Winston.
Chance, M. R. A. (1967). Attention structure as the basis of primate rank orders. *Man* (N.S.) **2**:503–18.
(1975). Social cohesion and the structure of attention. In *Biosocial Anthropology*, ed. R. Fox, pp. 93–113. London, Malaby Press.
(1976). Social attention: society and mentality. In *The Social Structure of Attention*, ed. M. R. A. Chance and R. R. Larsen, pp. 315–33. London, Wiley.
(1984). Biological systems synthesis of mentality and the nature of the two modes of mental operation: hedonic and agonic. *Man-Environmental Systems*, **14**:143–57.
(1988). A systems synthesis of mentality. In *Social Fabrics of the Mind* ed. M. R. A. Chance, pp. 37–45. Hove Erlbaum.
Chance, M. R. A. and Jolly, C. J. (1970). *Social Groups of Monkeys, Apes and Men*. London, Jonathan Cape.
Chance, M. R. A. and Larsen, R. R. (eds.) (1976). *The Social Structure of Attention*. London, John Wiley and Sons.
Chase, I. D. (1974). Models of hierarchy formation in animal societies. *Behaviorial Science* **19**:374–82.
(1980). Social process and hierarchical formation in small groups: a comparative perspective. *American Sociological Review* **45**:905–24.

Clark, C. B. (1977). A preliminary report on weaning among chimpanzees of the Gombe National Park, Tanzania. In *Primate Bio-social Development: Biological, Social and Ecological Determinants*, ed. S. Chevalier-Skolnikoff and F. Poirier, pp. 235–60. New York, Garland.

Crook, J. H. (1986). An undaunted 'starer at animals'. *New York Times Book Review*, 24 August: 1–23.

Darwin, C. (1871). *The Descent of Man, and Selection in Relation to Sex*, Part 2. New York, Appleton.

de Waal, F. *see* Waal, F. de

Deag, J. (1980). *Social Behaviour of Animals*. London, Edward Arnold.

Dittus, W. P. J. (1975). Discussion. In *Socioecology and Psychology of Primates*, ed. R. Tuttle, pp. 231–42. The Hague, Mouton.

Draper, P. (1973). Crowding among hunter–gatherers: the !Kung Bushmen. *Science* **182**:301–3.

(1975). !Kung women: Contrasts in sexual egalitarianism in foraging and sedentary contexts. In *Toward an Anthropology of Women*, ed. R. R. Reiter pp. 77–109. New York, Monthly Review Press.

(1978). The learning environment for aggression and anti-social behavior among the !Kung. In *Learning Non-aggression*, ed. A. Montagu, pp. 31–53. New York, Oxford University Press.

Dubos, R. (1973). Man's nature and social institutions. In *Man and Aggression*, ed. A. Montagu, 2nd edn, pp. 84–91. New York, Oxford University Press.

Durkheim, E. (1953). *Sociology and Philosophy*. New York, The Free Press.

Endicott, K. L. (1979). *Batek Negrito Religion*. Oxford, Clarendon Press.

Festinger, L. (1957). *A Theory of Cognitive Dissonance*. Evanston IL, Row, Peterson.

(1962). Cognitive dissonance. *Scientific American* **207**:93–102.

Festinger, L., Pepitone, A. and Newcomb, T. (1952). Some consequences of de-individuation in a group. *Journal of Abnormal and Social Psychology* **47**:382–9.

Foster, G. M., Scudder, T. Coleson, E. and Kemper, R. V. (eds.) (1979). *Long-term Field Research in Social Anthropology*. New York, Academic Press.

Fried, M. H. (1967). *The Evolution of Political Society*. New York, Random House.

Gallup, G. G., Jr (1970). Chimpanzees: self-recognition. *Science*, **167**:86–7.

(1975). Towards an operational definition of self-awareness. In *Sociology and Psychology of Primates*, ed. R. H. Tuttle, pp. 309–41. The Hague, Mouton.

Garner, R. L. (1896). *Gorillas and Chimpanzees*. London, Osgood, McClaine & Co.

Gardner, P. M. (1968). Discussions, Part VII. In *Man the Hunter*, ed. R. B. Lee and I. DeVore, pp. 335–45. Chicago, Aldine.

(1972). The Paliyans. In *Hunter–Gathers Today*, ed. M. G. Bicchieri, pp. 404–7. New York, Holt, Rinehart & Winston.

Gartlan, J. S. (1968). Structure and function in primate society. *Folia Primatologica* **8**:89–120.

Ghiglieri, M. P. (1979). The socioecology of chimpanzees in Kibale Forest, Uganda. Ph.D thesis, University of California, Davis, CA.

(1984). *The Chimpanzees of Kibale Forest*: A Field Study of Ecology and Social Structure. New York, Columbia University Press.

(1985). The social ecology of chimpanzees. *Scientific American*, **252**:102–13.

(1988). *East of the Mountains of the Moon*. New York, The Free Press.

(1989). Hominoid sociobiology and hominid social evolution. In *Understanding Chimpanzees*, ed. P. Heltne and L. Marquardt, pp. 370–9. Cambridge MA, Harvard University Press.

Gibb, C. A. (1947). The principles and traits of leadership. *Journal of Abnormal and Social Psychology* **42**:267–84.

(1970*a*). An interactional view of the emergence of leadership. In *Leadership*, ed. C. A. Gibb, pp. 214–22. Harmondsworth, Middx, Penguin Books.

(1970*b*). The principles and traits of leadership. In *Leadership*, ed. C. A. Gibb, pp. 205–13. Harmondsworth, Middx, Penguin Books.

Goffman, E. (1956). The nature of deference and demeanor. *American Anthropologist* **58**:473–502.

(1971). *Relations in Public*. New York, Harper and Row.

Goodall, J. (1963*a*). Feeding behaviour of wild chimpanzees: a preliminary report. *Symposia of the Zoological Society of London* **10**:39–48.

(1963*b*). My life among wild chimpanzees. *National Geographic* **124**:272–308.

(1965*a*). Chimpanzees of the Gombe Stream Reserve. In *Primate Behavior*, ed. I. DeVore, pp. 425–73. New York, Holt, Rinehart & Winston.

(1965*b*). New discoveries among Africa's chimpanzees. *National Geographic* **128**:802–31.

(1967*a*). Mother–offspring relationships in free-ranging chimpanzees. In *Primate Ethology*, ed. D. Morris, pp. 287–346. London, Weidenfeld and Nicolson.

(1967*b*). *My Friends, the Wild Chimpanzees*. Washington DC, National Geographic Society.

(1968*a*). The behaviour of free-living chimpanzees in the Gombe Stream Reserve. *Animal Behaviour Monographs* **1**:165–311.

(1968*b*). A preliminary report on expressive movements and communication in the Gombe Stream Chimpanzees. In *Primates: Studies in Adaptation and Variability*, ed. P. Jay, pp. 313–74. New York, Holt, Rinehart & Winston.

(1971*a*). *In the Shadow of Man*. Glasgow, Collins.

(1971*b*). Some aspects of aggressive behaviour in a group of free-living chimpanzees. *International Social Science Journal* **23**(1):89–97.

(1973). The behaviour of chimpanzees in their natural habitat. *American Journal of Psychiatry* **130**:1–12.

(1975). The chimpanzee. In *The Quest for Man*, ed. V. Goodall, pp. 130–69. London, Phaidon.

(1977). Infant killing and cannibalism in free-living chimpanzees. *Folia Primatologica* **28**:259–82.

(1979). Life and death at Gombe. *National Geographic* **155**:592–621.

(1983*a*). Order without law. In *Law, Biology and Culture*, ed. M. Gruter and P. Bohannan, pp. 50–62. Santa Barbara, CA, Ross-Erikson.
(1983*b*). Population dynamics during a 15 year period in one community of free-living chimpanzees in the Gombe National Park, Tanzania. *Zeitschrift für Tierpsychologie* **61**:1–60.
(1986*a*). Social rejection, exclusion and shunning among the Gombe chimpanzees. *Ethology and Sociobiology* **7** :227–36.
(1986*b*). *The Chimpanzees of Gombe: Patterns of Behavior*. Cambridge, MA, Harvard University Press.
(1990). *Through a Window*. Boston, MA, Houghton Mifflin.
Goodall, J., Bandora, A. Bergmann, E. Busse, C. Matama, H. Mpongo, E. Pierce, A. and Riss, D. (1979). Intercommunity interactions in the chimpanzee population of the Gombe National Park. In *The Great Apes*, ed. D. A. Hamburg and E. R. McCown, pp. 13–53. Menlo Park, CA, Benjamin/Cummings.
Hall, C. S. (1941). Temperament: a survey of animal studies. *Psychology Bulletin* **38**:909–43.
Hall, K. R. L. and DeVore, I. (1965). Baboon social behavior. In *Primate Behavior*, ed. I. DeVore, pp. 53–110. New York, Holt, Rinehart and Winston.
Hamburg, D. A. (1963). Emotions in the perspective of human evolution. In *Expression of the Emotions in Man*, ed. P. H. Knapp, pp. 300–17. New York, International Universities Press Inc.
Hamburg, D. A. Hamburg, B. A. and Barchas, J. D. (1975). Anger and depression in perspective of behavioral biology. In *Emotions – Their Parameters and Measurement*, ed. L. Levi, pp. 235–78. New York, Raven Press.
Hamilton, M. (1982). Symptoms and assessment of depression. In *Handbook of Affective Disorders*, ed. E. S. Paykel, pp. 3–11. New York, Guilford Press.
Hamilton, W. D. (1964). The genetic evolution of social behaviour, I, II. *Journal of Theoretical Biology* **7**:1–52.
(1972). Altruism and related phenomena, mainly in social insects. *Annual Review of Ecology and Systematics* **3**:193–232.
Harako, R. (1981). The cultural ecology of hunting behaviour among Mbuti pygmies in the Ituri Forest, Zaire. In *Omnivorous Primates*, ed. R. S. O. Harding and G. Teleki, pp. 449–555. New York, Columbia University Press.
Hayaki, H. (1985). Copulation of adolescent male chimpanzees, with special reference to the influence of adult males, in the Mahale National Park, Tanzania. *Folia Primatologica* **44**:148–60.
Hebb, D. O. (1946). Emotion in man and animal: an analysis of the intuitive processes of recognition. *Psychological Review* **53**:88–106.
Hinde, R. A. (1972). Aggression. In *Biology and the Human Sciences*, ed. J. W. S. Pringle, pp. 1–24. Oxford, Clarendon Press.
(1986). Can nonhuman primates help us understand human behavior? In *Primate Societies*, ed. B. B. Smuts, D. L. Cheney, R. M. Seyfarth, R. W. Wrangham and T. T. Struhsaker, pp. 413–20. Chicago, University of Chicago Press.

(1987). *Individuals, Relationships and Culture*. Cambridge, Cambridge University Press.
Hinde, R. A. and Stevenson-Hinde, J. (1976). Towards understanding relationships: dynamic stability. In *Growing Points in Ethology*, ed. P. P. G. Bateson and R. A. Hinde, pp. 451–79. Cambridge, Cambridge University Press.
Hiraiwa-Hasegawa, M., Hasegawa, T. and Nishida, T. (1984). Demographic study of a large-sized unit-group of chimpanzees in the Mahale Mountains, Tanzania: a preliminary report. *Primates* **25**(4):401–13.
Hold, B. C. L. (1976). Attention structure and rank specific behaviour in preschool children. In *The Social Structure of Attention*, ed. M. R. A. Chance and R. Larsen, pp. 177–201. London, John Wiley & Sons.
Hollander, E. P. (1970). Emergent leadership and social influence. In *Leadership*, ed. C. A. Gibb, pp. 293–306. Harmondsworth, Middx, Penguin Books.
Homans, G. C. (1950). *The Human Group*. New York, Harcourt Bruce.
Howell, N. (1979). *Demography of the Dobe !Kung*. New York, Academic Press.
Hrdy, S. (1976). The care and exploitation of nonhuman primate infants by conspecifics other than the mother. In *Advances in the Study of Behavior*, vol. 6, ed. J. Rosenblatt, R. Hinde, E. Shaw and C. Beer, pp. 101–58. New York, Academic.
(1979). Infanticide among animals: a review, classification and examination of the implications for the reproductive strategies of females. *Ethology and Sociobiology* **1**:13–40.
(1981). *The Woman that Never Evolved*. Cambridge, MA, Harvard University Press.
Itani, J. (1972). A preliminary essay on the relationship between social organization and incest avoidance in nonhuman primates. In *Primate Socialization*, ed. F. Poirier, pp. 165–71. New York, Random House.
(1979). Distribution and adaptation of chimpanzees in an arid area. In *The Great Apes*, eds. D. A. Hamburg and E. R. McCown, pp. 55–71. Menlo Park, CA, Benjamin/Cummings.
(1982). Intraspecific killing among non-human primates. *Journal of Social and Biological Structures* **5**:361–8.
Itani, J. and Suzuki, A. (1967). The social unit of chimpanzees. *Primates* **8**:355–81.
Izard, C. E. (1972). *Patterns of Emotion*. New York, Academic Press.
Izard, C. E. and Tompkins, S. S. (1966). Affect and behaviour: anxiety as a negative affect. In *Anxiety and Behavior*, ed. C. D. Spielberger, pp. 81–125. New York, Academic Press.
Izawa, K. (1970). Unit groups of chimpanzees and their nomadism in the savanna woodland. *Primates* **11**:1–46.
Izawa, K. and Itani, J. (1966). Chimpanzees in Kasakati Basin, Tanganyika. 1. Ecological study in the rainy season 1963–1964. *Kyoto University African Studies* **1**:73–156.
Janis, I. L. (1971). *Stress and Frustration*. New York, Harcourt Brace Jovanovich Inc.

Jolly, A. (1972). *The Evolution of Primate Behavior*. New York, Macmillan.
Kano, T. (1971). The chimpanzees of Filabanga, Western Tanzania. *Primates* **12**:229–46.
Kawabe, M. (1966). One observed case of hunting behavior among wild chimpanzees living in the savanna woodland of western Tanzania. *Primates* **7**: 393–6.
Kawanaka, K. (1981). Infanticide and cannibalism in chimpanzees – with special reference to the newly observed case in the Mahale Mountains. *African Studies Monographs* **1**: 69–99.
Kemper, T. D. (1978). Toward a sociology of emotions. *The American Sociologist* **13**: 30–41.
Kortlandt, A. (1962). Chimpanzees in the wild. *Scientific American* **206**(5): 128–38.
(1975). In discussion. In *Socioecology and Psychology of Primates*, ed. R. H. Tuttle, pp. 301–5. The Hague, Mouton.
Kortlandt, A. and Kooij, M. (1963). Protohominid behaviour in primates. *Symposium of the Zoological Society of London* **10**: 61–88.
Krebs, J. R. (1981). Foraging. In *The Oxford Companion to Animal Behaviour*, ed. D. McFarland, pp. 214–17. Oxford, Oxford University Press.
Krebs, J. R. and Davies, N. B. (eds.) (1978). *Behavioural Ecology*. Oxford, Blackwell.
(1981). *An Introduction to Behavioural Ecology*. Oxford, Blackwell.
Kuhn, T. S. (1970). *The Structure of Scientific Revolutions*, 2nd edn. Chicago, University of Chicago Press.
Kurland, J. A. and Beckerman, S. J. (1985). Optimal foraging and hominid evolution: labor and reciprocity. *American Anthropologist* **87**: 73–93.
Lancaster, J. (1986). Primate social behavior and ostracism. *Ethology and Sociobiology* **7**: 215–25.
Lawick-Goodall, J. van, *see* Goodall, J.
Leacock, E. and Lee, R. B. (eds.) (1982). *Politics and History in Band Societies*. Cambridge, Cambridge University Press.
Lee, R. B. (1968). What hunters do for a living, or, how to make out on scarce resources. In *Man the Hunter*, ed. R. B. Lee and I. DeVore, pp. 30–48. New York, Aldine.
(1979). *The !Kung San: Men, Women and Work in a Foraging Society*. Cambridge, Cambridge University Press.
(1984). *The Dobe !Kung*. New York, Holt, Rinehart & Winston.
Lee, R. B. and DeVore I. (eds.) (1968). *Man the Hunter*. New York, Aldine.
(1976). *Kalahari Hunter–gatherers: Studies of the !Kung San and Their Neighbours*. Cambridge, MA, Harvard University Press.
Levi-Strauss, C. (1968). The concept of primitiveness. In *Man the Hunter*, ed. R. B. Lee and I. DeVore, pp. 349–52. New York, Aldine.
McFarland, D. (1981). Displays. In *The Oxford Companion to Animal Behaviour*, ed. D. McFarland, pp. 133–6. New York, Oxford University Press.
McGinnis, P. R. (1979). Sexual behavior in free-living chimpanzees: consort relationships. In *The Great Apes*, ed. D. A. Hamburg and E. R. McCown, pp. 429–39. Menlo Park, CA, Benjamin/Cummings.

MacKinnon, J. (1978). *The Ape Within Us*. New York, Holt, Rinehart & Winston.
Mackal, P. K. (1979). *Psychological Theories of Aggression*. Amsterdam, North-Holland Publishing Co.
Maier, N. B. F. (1961). *Frustration: The Study of Behavior Without a Goal*. Ann Arbor, University of Michigan Press.
Marler, P. (1976). Social organization, communication and graded signals: the chimpanzee and the gorilla. In *Growing Points in Ethology*, ed. P. P. G. Bateson and R. A. Hinde, pp. 239–80. Cambridge, Cambridge University Press.
Marshall, L. (1957). The kin terminology system of the !Kung Bushmen. *Africa* **27**: 1–25.
(1960). !Kung Bushmen bands. *Africa* **30**: 325–55.
(1965). The !Kung Bushmen of the Kalahari Desert. In *Peoples of Africa*, ed. J. Gibb, pp. 241–78. New York, Holt, Rinehart & Winston.
(1976). *The !Kung of Nyae Nyae*. Cambridge, MA, Harvard University Press.
Mason, W. A. (1964). Sociability and social organization in monkeys and apes. In *Advances in Experimental Social Psychology 1*, ed. L. Berkowitz, pp. 277–305. New York, Academic Press.
Maynard Smith, J. and Price, G. R. (1973). The logic of animal conflict. *Nature* **246**: 15–18.
Mayr, E. (1970). *Populations, Species and Evolution*. Cambridge, MA, The Belknap Press.
Midgley, M. (1978). *Beast and Man: The Roots of Human Nature*. Ithaca, NY, Cornell University Press.
Milgram, S. (1974). *Obedience to Authority*. New York, Harper and Row.
Montagner, H., Henry, J., Lombardot, M., Restoin, A., Benedini, M., Godard, D., Boillot, F., Pretot, M., Bolzoni, D., Burnod, J. and Nicolas, R. (1979). The ontogeny of communication behaviour and adrenal physiology in the young child. *Child Abuse and Neglect* **3**: 19–30. Oxford, Pergamon Press.
Montagner, H., Henry, J., Lombardot, M., Restoin, A., Bolzoni, D., Durand, M., Humbert, Y. and Moyse, A. (1978). Behavioural profiles and corticosteroid excretion rhythms in young children. Part 1: Non-verbal communication and setting up of behavioural profiles in children from 1 to 6 years. In *Human Behaviour and Adaptation*, ed. N. Blurton-Jones and V. Reynolds, pp. 207–28. London, Taylor and Francis.
Montagu, A. (1966). *On Being Human*. New York, Hawthorn.
(1970). *The Direction of Human Development*, revised edn. New York, Hawthorn.
(1976). *The Nature of Human Aggression*. New York, Oxford University Press.
(1978). *Learning Non-Aggression*. New York, Oxford University Press.
(1987). There is no human propensity to kill. *Letters, The New York Times*: 7 June.

Mori, A. (1982). An ethological study on chimpanzees at the artificial feeding place in Mahale Mountains, Tanzania – with special reference to the booming situation. *Primates* **23**: 45–65.
Moyer, K. E. (1976). *The Psychobiology of Aggression*. New York, Harper and Row.
Murdock, G. P. (1968). The current status of the world's hunting and gathering peoples. In *Man the Hunter*, ed. R. B. Lee and I. DeVore, pp. 13–20. Chicago, Aldine.
Myers, J. P., Connors, P. G. and Pitelka, F. A. (1981). Optimal territory size and the sanderling: compromises in a variable environment. In *Foraging Behavior*, ed. A. C. Kamil and T. D. Sargent, pp. 135–58. New York, Garland.
Nelson, L. L. and Kagan, S. (1972). Competition, the star-spangled scramble. *Psychology Today*, September: 53–6, 90–1.
Nicolson, N. A. (1977). A comparison of early behavioral development in wild and captive chimpanzees. In *Primate Bio-social Development: Biological, Social and Ecological Determinants*, ed. S. Chevalier-Skolnikoff and F. E. Poirier, pp. 529–60. New York, Garland.
Nisbet, R. and Perrin, R. G. (1977). *The Social Bond*, 2nd edn. New York, Knopf.
Nishida, T. (1968). The social group of wild chimpanzees in the Mahali Mountains. *Primates* **9**: 167–224.
(1970). Social behavior and relationship among wild chimpanzees of the Mahali Mountains. *Primates* **11**: 47–87.
(1979). The social structure of chimpanzees of the Mahale Mountains. In *The Great Apes*, ed. D. A. Hamburg and E. R. McCown, pp. 73–121. Menlo Park, CA, Benjamin/Cummings.
(1980). The leaf-clipping display: a newly discovered expressive gesture in wild chimpanzees. *Journal of Human Evolution* **9**: 117–28.
(1983*a*). Alloparental behavior in wild chimpanzees of the Mahale Mountains, Tanzania. *Folia Primatologica* **41**: 1–33.
(1983*b*). Alpha status and agonistic alliance in wild chimpanzees. *Primates* **24**: 318–36.
Nishida, T. and Hiraiwa-Hasegawa, M. (1986). Chimpanzees and bonobos: cooperative relationship among males. In *Primate Societies*, ed. B. B. Smuts, D. Cheney, R. M. Seyfarth, R. Wrangham and T. T. Struksaker, pp. 165–77. Chicago, University of Chicago Press.
Nishida, T., Hiraiwa-Hasegawa, M. Hasegawa, T. and Takahata, Y. (1985). Group extinction and female transfer in wild chimpanzees in the Mahale National Park, Tanzania. *Zeitschrift für Tierpsychologie* **67**: 281–301.
Nishida, T. and Kawanaka, K. (1972). Inter-unit-group relationships among wild chimpanzees of the Mahale Mountains. *Kyoto University African Studies* **7**: 131–69.
(1985). Within-group cannibalism by adult male chimpanzees. *Primates* **26**: 274–84.
Nishida, T., Vehara, S. and Ramadhani, N. (1979). Predatory behavior among chimpanzees of the Mahale Mountains. *Primates* **20**: 1–20.

Nissen, H. W. (1931). A field study of the chimpanzee. *Comparative Psychology Monographs* **8**:1–22.
(1956). Individuality in behavior of chimpanzees. *American Anthropologist* **58**:407–13.
Norikoshi, I. (1982). One observed case of cannibalism among wild chimpanzees of the Mahale Mountains. *Primates* **23**:66–74.
Poirier, F. (1972). Introduction. In *Primate Socialization*, ed. F. Poirier, pp. 3–28. New York, Random House.
Power, M. (1986). The foraging adaptation of chimpanzees, and the recent behaviors of the provisioned apes in Gombe and Mahale National Parks, Tanzania. *Human Evolution* **1**:251–66.
(1988). The cohesive foragers: human and chimpanzee. In *Social Fabrics of the Mind*, ed. M. R. A. Chance, pp. 75–103. Hove, Erlbaum.
Pusey, A. (1979). Intercommunity transfer of chimpanzees in Gombe National Park. *The Great Apes*, ed. D. A. Hamburg and E. R. McCown, pp. 465–79. Menlo Park, CA, Benjamin/Cummings.
Pusey, A. and Packer, C. (1986). Dispersal and philopatry. In *Primate Societies*, ed. B. B. Smuts, D. Cheney, R. Seyfarth, R. Wrangham and T. Struhsaker, pp. 250–66. Chicago, University of Chicago Press.
Ransom, T. W. (1981). *Beach Troop of the Gombe*. Lewisburg, PA, Bucknell University Press.
Reynolds, V. (1963). An outline of the behaviour and social organisation of forest-living chimpanzees. *Folia Primatologia* **1**:95–102.
(1964). The man of the woods. *Natural History* **73**:44–51.
(1965*a*). *Budongo: An African Forest and Its Chimpanzees*. New York, The Natural History Press.
(1965*b*). Some behavioral comparisons between the chimpanzees and the mountain gorilla in the wild. *American Anthropologist* **67**:691–706.
(1966). Open groups in hominid evolution. *Man* (N.S.) **1**:441–52.
(1968). Kinship and the family in monkeys, apes and man. *Man* (N.S.) **3**:209–23.
(1970). Roles and role change in monkey society: the consort relationship of rhesus monkeys. *Man* (N.S.) **5**:449–65.
(1975). How wild are the Gombe chimpanzees? *Man* (N.S.) **10**:123–5.
(1980). *The Biology of Human Action*, 2nd edn. Oxford, W. H. Freeman.
Reynolds, V. and Reynolds F. (1965). Chimpanzees of the Budongo Forest. In *Primate Behavior*, ed. I. DeVore, pp. 368–424. New York, Holt, Rinehart & Winston.
Riss, D. and Busse, C. (1977). Fifty-day observation of a free-ranging adult male chimpanzee. *Folia Primatologica* **28**:283–97.
Rogers, C. M. and Davenport, R. K. (1970). Chimpanzee maternal behavior. In *The Chimpanzee*, ed. G. H. Bourne, vol. 3, pp. 361–8. Basel, Karger.
Rousseau, J. J. (1762). *Du Contrat Social*; transl. G. D. H. Cole, 1938, *The Social Contract*. New York, Dutton.
Rowell, T. E. (1972). *The Social Behaviour of Monkeys*. Harmondsworth, Middx, Penguin Books.
(1974). The concept of social dominance. *Behavioral Biology* **11**:131–54.

Rowell, T. E. and Olson, D. K. (1983). Alternative mechanisms of social organization in monkeys. *Behaviour* **86**:31–54.

Russell, C. and Russell, W. M. S. (1968). *Violence, Monkeys and Man.* London, Macmillan.

(1969). The natural history of violence. *Journal of Medical Ethics* **5**:108–17.

(1972). Primate male behavior and its human analogues. In *Primates on Primates*, ed. D. D. Quiatt, pp. 47–58. Boulder, CO, University of Colorado Press.

Saayman, G. S. (1975). The influence of hormonal and ecological factors upon sexual behavior and social organization in old world primates. In *Socioecology and Psychology of Primates,* ed. R. Tuttle, pp. 181–204. The Hague, Mouton.

Saayman, G. S. and Taylor, C. K. (1973). Social organization of inland dolphins (Tursiops Aduncus and Sousa) in the Indian Ocean. *Journal of Mammalogy* **54**:993–6.

Sade, D. S. (1972). A longitudinal study of social behavior of rhesus monkeys. In *The Functional and Evolutionary Biology of Primates*, ed. R. Tuttle, pp. 378–98. The Hague, Mouton..

Schachter, S. (1959). *The Psychology of Affiliation*. Stanford, CA, Stanford University Press.

Selye, H. (1974). *Stress Without Distress*. New York, Lippincott.

Service, E. R. (1962). *Primitive Social Organization: An Evolutionary Perspective*. New York, Random House.

(1966). *The Hunters*. Englewood Cliffs, N.J., Prentice Hall.

Shafton, A. (1976). *Conditions of Awareness: Subjective Factors in the Social Adaptations of Man and Other Primates*. Portland, OR, Riverstone.

Shils, E. (1965). Charisma, order and status. *American Sociological Review* **30**:199–213.

Silberbauer, G. (1972). The G/wi Bushmen. In *Hunters and Gatherers Today*, ed. M. Bicchieri, pp. 271–325. New York, Holt, Rinehart & Winston.

(1981). Hunter/gatherers of the central Kalahari. In *Omnivorous Primates: Gathering and Hunting in Human Evolution*, ed. R. Harding and G. Teleki pp. 455–98. New York, Columbia University Press.

(1982). Political process in G/wi bands. In *Politics and History in Band Societies*, ed. E. Leacock. and R. Lee, pp. 23–35. London, Cambridge University Press.

Simpson, M. J. A. (1973). The social grooming of male chimpanzees: a study of eleven free-living males in the Gombe Stream National Park, Tanzania. In *Comparative Ecology and Behavior of Primates*, ed. R. P. Michael and J. H. Crook, pp. 411–505. New York, Academic Press.

Sorenson, E. R. (1978). Cooperation and freedom among the Fore of New Guinea. In *Learning Non-Aggression* ed. A. Montagu, pp. 12–30. Oxford, Oxford University Press.

Staub, E. (1984). Notes toward an interactionist-motivational theory of the determinants and development of (pro)social behavior. In *Development and Maintenance of Prosocial Behavior*, ed. E. Staub, D. Bar-Tal, J. Karylowski and J. Reykowski. pp. 29–49. New York, Plenum Press.

Steklis, H. D. (1985). Primate communication and comparative neurology, and the origin of language re-examined. *Journal of Human Evolution* **14**:157–73.

Strum, S. C. (1987). *Almost Human*. New York, Random House.

Sugiyama, Y. (1968). Social organization of chimpanzees in the Budongo Forest, Uganda. *Primates* **9**:225–58.

(1969). Social behavior of chimpanzees in the Budongo Forest, Uganda. *Primates* **10**:197–225.

(1972). Social characteristics and socialization of wild chimpanzees. In *Primate Socialization*, ed. F. E. Poirer, pp. 145–63. New York, Random House.

(1973). Social organization of wild chimpanzees. In *Behavioral Regulators of Behavior in Primates* ed. C. R. Carpenter, pp. 68–80. Lewisburg, PA, Bucknell University Press.

(1984). Population dynamics of wild chimpanzees at Bossou, Guinea, between 1976 and 1983. *Primates* **25**:391–400.

(1989). Populations dynamics of chimpanzees at Bossou, Guinea. In *Understanding Chimpanzees*, eds. P. Heltne and L. G. Marquardt, pp. 134–45. Cambridge, MA, Harvard University Press.

Sugiyama, Y. and Koman, J. (1979). Social structure and dynamics of wild chimpanzees at Bossou, Guinea. *Primates* **20**:323–39.

Suzuki, A. (1969). An ecological study of chimpanzees in a savanna woodland. *Primates* **10**:103–48.

(1971). Carnivority and cannibalism observed among forest-living chimpanzees. *Journal of the Anthropological Society of Nippon* **79**:30–48.

(1975). The origin of hominid hunting: a primatological perspective, In *Socioecology and Psychology of Primates*, ed. R. H. Tuttle, pp. 259–78. The Hague, Mouton.

Takahata, Y. (1985). Adult male chimpanzees kill and eat a male newborn infant: newly observed intragroup infanticide and cannibalism in Mahale National Park, Tanzania. *Folia Primatologica* **44**:161–70.

Tanaka, J. (1980). *The San*, hunter-gatherers of the Kalahari. Tokyo. University of Tokyo Press.

Teleki, G. (1973). *The Predatory Behavior of Wild Chimpanzees*. Lewisburg, PA, Bucknell University Press.

(1975). Primate subsistence patterns: collector–predators and gatherer–hunters. *Journal of Human Evolution* **4**:125–84.

(1981). The omnivorous diet and eclectic feeding habits of chimpanzees in Gombe National Park, Tanzania: In *Omnivorous Primates, Gathering and Hunting in Human Evolution*, ed. R. S. O. Harding and G. Teleki, pp. 303–43. New York, Columbia University Press.

Teleki, G., Hunt, E. E., Jr and Pfifferling, J. H. (1976). Demographic observations (1963–1973) on the chimpanzees of Gombe National Park, Tanzania. *Journal of Human Evolution* **5**:559–98.

Theodorson, G. A. and Theodorson, A. G. (1969). *A Modern Dictionary of Sociology*. New York, Crowell.

Thibaut, J. W. and Kelley, H. H. (1959). *The Social Psychology of Groups*. New York, John Wiley and Sons.

Thomas, E. M. (1959). *The Harmless People*. New York, Knopf.

Tiger, L. (1969). *Men in Groups*. New York, Random House.

Turnbull, C. M. (1961). *The Forest People: A Study of the Pygmies of the Congo*. New York, Simon and Schuster.

(1965). The Mbuti Pygmies of the Congo. In *Peoples of Africa*, ed. J. L. Gibbs, pp. 279–317. New York, Holt, Rinehart & Winston.

(1966). *Wayward Servants: The Two Worlds of the African Pygmies*. London, Eyre and Spottiswoode.

(1968*a*). Contemporary societies: the hunters. *International Encyclopedia of the Social Sciences* **7**:21–6.

(1968*b*). Discussions, Part VII. Primate behavior and the evolution of aggression. In *Man the Hunter*, ed. R. B. Lee and I. DeVore pp. 339–44. Chicago, Aldine.

(1972). Demography of small scale societies. In *The Structure of Human Populations* eds. G. A. Harrison and A. J. Boyce, pp. 283–312. Oxford, Clarendon Press.

(1978). The politics of non-aggression. In *Learning Non-Aggression*, ed. A. Montagu, pp. 161–221. New York, Oxford University Press.

(1983). *The Human Cycle*. New York, Simon and Schuster.

Tutin, C. E. G. (1975). Exceptions to promiscuity in a feral chimpanzee community. In *Contemporary Primatology*, Proceedings of the Fifth International Congress of Primatology, ed. S. Kondo, M. Kawai and A. Ehara, pp. 445–9. Basel, Karger.

(1979). Mating patterns and reproductive strategies in a community of wild chimpanzees (*Pan troglodytes schweinfurthii*). *Behavioral Ecology and Sociobiology* **6**:29–38.

Tutin, C. E. G. and McGinnis, P. R. (1981). Chimpanzee reproduction in the wild. In *Reproductive Biology of the Great Apes*, ed. C. E. Graham, pp. 239–64. New York, Academic Press.

Uehara, S. and Nyundo, R. (1983). One observed case of temporary adoption of an infant by unrelated nulliparous females among wild chimpanzees in the Mahale Mountains, Tanzania. *Primates* **24**:456–66.

Vayda, A. P. (1976). *War in Ecological Perspective: Persistence, Change and Adaptive Process in Three Oceanian Societies*. New York, Plenum.

Waal, F. de (1982). *Chimpanzee Politics: Power and Sex Among Apes*. New York, Harper and Row.

(1986). The brutal elimination of a rival among captive male chimpanzees. *Ethology and Sociobiology* **7**:237–51.

(1989). *Peacemaking among Primates*. Cambridge, MA, Harvard University Press.

Waal, F. de and Hoekstra, J. A. (1980). Contexts and predictability of aggression in chimpanzees. *Animal Behaviour* **28**:929–37.

Waal, F. de and Roosmalen, A. van (1979). Reconciliation and consolation among chimpanzees. *Behavioral Ecology and Sociobiology* **5**:55–66.

Washburn, S. L. (1980). Human behavior and the behavior of other animals. In *Sociobiology Examined* ed. A. Montagu, pp. 254–82. Oxford, Oxford University Press.

Weber, M. (1957). *The Theory of Social and Economic Organization*. Glencoe, IL, The Free Press.

Wedgewood–Oppenheim, F. (1988). Organisational culture and the agonic/hedonic bimodality. In *Social Fabrics of the Mind*, ed. M. R. A. Chance, pp. 312–20. Hove, Erlbaum.

Woodburn, J. (1968*a*). An introduction to Hadza economy. In *Man the Hunter*, ed. E. Lee and I. DeVore, pp. 49–55. Chicago, Aldine.

(1968*b*). Stability and flexibility in Hadza residential groupings. In *Man the Hunter*, ed. R. Lee and I. DeVore, pp. 103–10. Chicago, Aldine.

(1970). *Hunters and Gatherers: The Material Culture of the Nomadic Hadza*. London, British Museum.

(1972). Ecology, nomadic movement and the composition of the local group among hunters and gatherers. In *Man, Settlement and Urbanism* ed. P. Urko, R. Tringham and G. Dimbleby, pp. 193–206. London, Duckworth.

(1979). Minimal politics: the political organization of Hadza of North Tanzania. In *Politics in Leadership*, ed. W. A. Shack and P. S. Cohen, pp. 244–66. Oxford, Clarendon Press.

(1980). Hunters and gatherers today and reconstruction of the past. In *Soviet and Western Anthropology*, ed. E. Gellner, pp. 95–117. London, Duckworth.

(1982). Egalitarian societies. *Man* (N.S.) **17**:431–51.

Wrangham, R. W. (1974). Artificial feeding of chimpanzees and baboons in their natural habitat. *Animal Behaviour* **22**:83–93.

(1977). Feeding behaviour of chimpanzees in Gombe National Park, Tanzania. In *Primate Ecology: Studies of Feeding and Ranging Behaviour in Lemurs, Monkeys and Apes*, ed. T. H. Clutton Brock, pp. 503–38. London, Academic Press.

(1979). Sex differences in chimpanzee dispersion. In *The Great Apes*, ed. D. A. Hamburg and E. R. McCown, pp. 481–90. Menlo Park, CA., Benjamin/Cummings.

(1982). Mutualism, kinship and social evolution. In *Current Problems in Sociobiology* ed. King's College Sociobiology Group, pp. 269–89. Cambridge, Cambridge University Press.

Yellen, J. E. (1976). Settlement patterns of the !Kung: An archeological perspective. In *Kalahari Hunter-Gatherers*, ed. R. B. Lee and I. DeVore, pp. 47–72. Cambridge, MA, Harvard University Press.

Yerkes, R. M. and Elder, J. H. (1936). Oestrus, receptivity and mating in chimpanzees. *Comparative Psychology Monographs* **13**(5):1–39.

Zimbardo, P. G. and Ruch, F. L. (eds.) (1975). *Psychology and Life*. 9th edn. Glenview, IL, Scott, Foresman.

Zuckerman, S. (1932). *The Social Life of Monkeys and Apes*. London, Routledge and Kegan Paul.

INDEX

adaptation
  aggressive behavior, when adaptive 14–15, 239–44, 248
  animal 239–40
  basic human organizational 37
  chimpanzee 241
  for reproductive success 95
adolescence, period of
  culturally defined 203
  defined 117
  experience of provisioned chimpanzees 105–22, 205, 209, 219–20
  missing, foraging society 117
  missing, wild chimpanzee groups 117
  psychobiology of adolescent anger 119
  see also youth, period of
adult rank, achievement of
  by foragers 75–6, 103, 117–22 *passim*, 203–5
  in provisioned groups 75–6, 103, 117–22
  social and maturational factors in 122, 203–5
  sole rank, immediate-return society 41, 46
  by wild chimpanzees 203–24
aggression
  when adaptive 14–15, 239–44, 248
  direct competition and 143, 144
  factors affecting arousal 75, 93, 137–9
  ideal learning situation 93
  maladaptive 140–2
  physiological expression of 69
  psychological basis for 31
  response to frustration 136–42
aggressive behavior
  current understandings based on 2, 3, 73, 88, 102–3, 112, 144–5, 147, 198, 206, 257
  disturbed foragers 14–15
  lack of, foraging societies 18–19, 47–8, 49
  lack of, wild chimpanzees 22–3, 25, 59, 66, 74, 118, 145, 159–61, 226, 260, 263
  genetic potential for 149, 239
  see also behavior, social *and* frustration
alliances, chimpanzees 122–3
alloparenting, *see* aunting
Amber, female chimpanzee, Arnhem zoo 198–9, 215
anger
  frustration-induced 140
  the psychobiology of 119
animal conflict, nature of 71
apprenticeship
  of chimpanzees 117–19, 120, 122, 205–9
  defined 117
  of youthful foragers 117, 203–5, 259
  see also adult rank
Arnhem zoo colony
  conditions 13, 255
  inability to utilize fission–fusion pattern 198, 199, 255
  unnatural stress of 13
  value of study 13, 197
artificial feeding, *see* provisioning
assertive behavior, 76, 139–40, 246
attachment to group
  necessity of, fission–fusion societies 217–43
  positive social contacts strengthen 187, 188

attachment to group *cont.*
social kinship 45–6
through carnival 230–1
see also social bonds
attention structure
agonic mode 189, 264–5
breakdown of 241
hedonic mode 189, 264–5
passing information by means of 188–93, 205, 208, 223
attraction
affiliation theory 231
attachment figures 172
basis of chimpanzee subgroups 62–3
charismatic and dependent chimpanzees 188
hedonic mode of 189
to larger society 134–5, 230
of mobile and sedentary groups 42–3, 232–3
youthful apes to males 120, 157, 206, 210–11
aunting 214–18
defined 214
female form of display 217–18
function of 215–17
authority
in foraging society 41, 46–8, 155, 170–1, 174–8, 198–9, 234–7, 247, 261
in mutual dependence system 152–5, 171–2
in wild chimpanzee groups 158–61, 168–70, 198, 199, 244–7, 265
theory 152–3, 247
autonomy
defined 63
of foragers 41, 47–8, 172, 197
principle of immediate-return system 41, 63
properties of 41, 63, 174, 250
theory 88
of wild chimpanzees 62, 63, 64–5, 129, 201, 206

baboons
charismatic leaders 266
mother–infant relations 259
relations with provisioned chimpanzees 28, 29–30, 59
relations with wild chimpanzees 28–30, 33–4, 58–9, 69–70
social organization, wild 262
social reciprocity 262
Basuta, I. 257–8
Batek Negritos, Malaysia
extant foraging people 40
behavior, social
affect connected with 8
factors in individual 84, 150–1, 201, 236
flexibility of, chimpanzee 248
flexibility of, foragers 42
genetic sets of 67, 140, 142, 149–50, 188, 240, 241–2, 264
negative 8, 67, 140, 150, 240, 248
positive 4, 8, 20, 41, 135, 187
role-related 179–80
see also aggressive behavior
behavior, ritualized
arena-related 231
of chimpanzees 81, 133, 160, 222–3, 227–8
deference 172–4
presentational rituals 174
Bernard, C. xiii
birth rate
adapted to normal mortality 95–6
among chimpanzees 95–6
among foragers 110

cannibalism 72, 95–103
abnormal behavior during 97–101, 102–3
explanations offered 101–3
by Gombe and Mahale chimpanzees 96–7, 99–102
rare among mammals 97
related to negative excitement 98–9, 103
in wild group 97–8
see also infanticide
carnival reunions
in business firms 252–3
in foraging society 227
restricted, Gombe 134, 227, 229, 260, 264
social atmosphere 63–5, 132, 133–4, 225–6, 227, 229–30, 232
social function of 132–5, 224–32, 239
wild chimpanzees 97, 132, 133–4, 224–32
see also larger society gatherings, visiting
charisma
defined 16
and dependence 16–17
extraordinary form of 165
normal form of 16, 165–6, 167
charismatic–dependent role relationship
balance of power in 174, 177, 250
in business management teams 252–3, 265

functions of 177
generalized parent–child relationship 155, 174–8, 261
fluidity of roles 9, 46–7, 155, 168, 170–1, 172, 245
of human children 251–2
of mobile and sedentary groups 10, 234
principles of 174–9, 250
rewards of 262
see also friendship
charismatic leaders
both sexes 47, 197, 198, 200, 261
elders 173, 234–7
extraordinary, chimpanzees 165–6, 169, 198–9, 200, 235–6, 246–7, 265
in foraging society 46–8
influential, without formal power 46, 47, 198, 199
potentially authoritarian 247–8
role of 153, 154, 155, 168, 199, 261
selected by group 47, 168, 173–4, 199, 246
traits of, chimpanzee 166–70, 208–9, 266
traits of, foraging society 170–1, 246, 261
see also leadership role
chimpanzees
attributes shared with humans 3–4, 7, 39, 84–5, 151, 161, 187, 188, 211, 239, 248
classification by age 84–5, 107, 113
communication 189, 190–3, 251
composition of wild groups 55–7, 60, 62–5, 74
genetic endowment 19, 162
ranges curtailed by human settlement 58, 248
relations with predators 59–60
sociability of wild 146
study of females neglected 196
wild and provisioned, defined 11–12
chimpanzees, female experience
Arnhem zoo 197–200, 264
current understandings of 106, 116, 121, 196, 197–8, 201–2, 212
provisioned groups 79–80, 94, 105–6, 116, 121–2, 129, 131, 200, 201, 202, 219, 220, 257, 262
wild groups 51–7, 77–9, 86, 107, 174–6, 151–239
cognitive dissonance theory 264
competition
of animals 149
Darwinian 251
modes of 15–16
resource 242–3
theory 143–50
competition, direct
absent among foragers 18, 47, 143
defined 15
and dominance theory 30
factors leading to 67
possessiveness 67
of provisioned and caged chimpanzees 16, 28–31, 67, 70, 73, 82, 104, 135, 138–9, 143–5, 241–2
tension of 16, 149, 241, 242, 259
theory 67, 73, 143–50
competition, indirect
absence of possessiveness 28
adapted mode of chimpanzees 16, 33, 73, 131, 142, 149, 241, 242
defined 16, 33
conformity
the independent conformists 179–81
psychological reward of 180–1
social control through 179–81
to social norms 180–1
crises 240–4
abnormal 255
behavior in acute food 243, 248
cultural 240–1
definitions 242, 255
natural 240, 255–6
role of endowed individuals in 247
simulated food 241, 243

Darwin, C. 47, 155, 158, 251
David Greybeard, male chimpanzee, Gombe 27, 163–4, 166, 167, 168, 169, 176–7, 205, 208, 246, 261, 262
deference
among chimpanzees 104, 159, 173, 174, 234–6
theory, the nature of 172–4
dependence
basis of 166
defined 16, 172
difficulty with concept 16–17, 172, 256
in immediate-return societies 172
negative form of 123
normal form of 17, 47, 88, 171–2
role of follower 171–2, 176, 177
despots
leadership theory excludes 167
rise of, in Gombe group 74–7, 167, 185, 246
submission to 82, 83, 123, 246–7, 265
display(s)
aggressive use of 75, 81, 91, 119, 121, 218, 230
aunting, female 217–18

display(s) *cont.*
  courtship 81
  functions of 75, 160, 218
  frustration 32
  goal of 81, 160, 218
  nonaggressive 64, 74, 75, 134, 225–6
dominance
  concept as aggressive, fixed 76
  defined 76
  equated with control 244–5
  function of 76
  paradigm 34–5, 196
  omitted from leadership theory 167
  see also hierarchy
dominance, aggressive
  absent, foraging societies 41, 46–8
  absent, wild chimpanzee groups 22–3, 25, 74
  factors mitigating against 42, 74, 132, 155, 177
  of provisioned and caged chimpanzees 19, 28–33, 74–7, 80, 103, 104, 119–22
  similar potential for, both sexes 198, 265

Durkheim, E. 163
egalitarian societies
  autonomy of adults 41, 47–8, 172, 195, 197
  balance of power in 171, 177
  characteristics of 41, 42
  extant, 37, 39–40
  immediate-return system 37–49, 87–8, 195–6
  mutual dependence system 8–9, 151–93
  principles of 18, 20, 40–3, 47, 48–9, 63, 131–2, 170, 172, 187, 193, 196
  see also foraging societies
egalitarianism
  in chimpanzee society 155, 195–237
  definitions of 41, 42, 195
  incompatibility with formal law 40–1, 43–4
  intensity of, foraging society 41, 195
  Rousseau's concept 248–51
  see also foraging societies
emotions
  bipolar nature of 7–8
  emotional disturbance in humans 110–11, 242
  emotional feelings 8
  emotional response to stress, chimpanzees 93, 142, 257–9
  emotive rewards 184, 186–8
  negative 7–8, 137–9
  positive 7–8
  positive emotive responses 48
  transpecific 5, 7
equilibrium
  dynamic 66, 177
  of foraging societies 48–9
  social mechanisms maintain 48
  theory 66
  of wild chimpanzee groups 66
excitement
  in cannibalism 97, 98
  in carnival gatherings 64, 133, 134, 229, 230, 232
  excitability thresholds 162
  negative and positive forms 147, 229–30
  of patrols 229–30
  psychological gratification of 64, 230
  raises rate of copulation 79, 257
exigency, category of 255

fathering role
  biological and social, foraging societies 158, 160–1
  generalized 157
  investment of male chimpanzees 156–8, 159–61
  social fathers, primates 85–92 *passim*, 157–61 *passim*, 175–6
  under stress 94
  see also parental roles
field studies
  categories for 12–13, 30
  different sets of evidence, naturalistic and provisioning 3, 5–6, 15
  ideal method for primate 21
  value of long term 67
  value in use of early 14–15
field studies, naturalistic
  basis for mutual dependence argument 5–6
  defined 12
  difficulty of 26–7
  Goodall's, pre-1965 29
  long term 21
  Mahale, pre-1968 30
  researchers and methods 21–8, 258–9
  tendency to undervalue 1–2, 19, 21, 23, 58, 256
  unanimity of reports 1, 2, 3, 65–6, 142, 256
  value of 77–8
field studies, provisioned chimpanzees
  advantages of 1, 6, 12
  broad use of term 12
  defined 12
  feeding methods 28–35
  Gombe, post-1965 29, 30

Mahale, post-1968 30
means of analysis, human-based theory 5
over-reliance on 14, 34–5
value of 6, 254
field studies, undisturbed foragers
early studies 14, 18, 19, 47, 84, 172, 230
immediate-return societies 39–40, 41–9 *passim*
value in reliance on early 14–15
Fifi, female chimpanzee, Gombe 107, 159, 172, 221
fission–fusion pattern
attenuated, provisioned groups 131–3
of chimpanzees 60–5, 224–32
disrupted by provisioning 29
fission aspect of 20–1, 132, 186, 231–2, 255–6
in foraging society 37, 42–5, 132
function of 44, 131–2, 231–2, 243
fusion aspect of 132
incompatibility with rigid hierarchy 132
operates all levels of chimpanzee society 132, 229
pata monkeys 190
fitness
of chimpanzees, reduced 130
Darwinian 155
foraging strategy and 125, 130
inclusive 54, 148, 217
Flint, male chimpanzee, Gombe 92–3, 107–9
Flo, female chimpanzee, Gombe 27, 107–9, 164–5, 219–20, 224, 233, 236, 246, 257
follower role 171–2
see also charismatic–dependent relationship *and* dependence
food, wild
availability of 58, 127, 129, 258
basic physiological need 33, 125–6, 138, 239–40
foraging strategy, chimpanzees 58, 125–50, 206, 207, 233, 241, 242
foraging strategy, foragers 40, 42–4
human interference with apes obtaining 28–33, 67–70, 241, 257–9
meat-eating, chimpanzees 98, 100
scattered nature of 43–4, 127, 130
food-calling of chimpanzees
benefit to society 133, 134
booming session 95
drumming 63, 64, 225
explanations of 228–9
as foraging strategy 133–4, 227
results in larger society carnival 134, 225–6, 227
foraging societies
anthropological understandings 18–19, 20–1, 37–49, 57, 87–8, 110, 132, 172, 174–5, 181, 196–7, 198, 199, 203–5, 233
attachment to group 46, 86–7, 131, 170, 177, 187, 217, 230, 243
effects of interference with 14–15, 18–19
generalized social parenting 87–8, 158, 160–1
individualization 195
leader and follower roles 46–7, 88, 155
negatively sanctioned traits 170
position of women in 195–6
positive characteristics of 18, 19, 20
self-esteem 188
values 131, 170
see also egalitarianism *and* egalitarian societies
foraging systems
delayed return 39
immediate-return 3–4, 20–1, 37, 39–43, 135
other models of 37
foraging theory
competition and gregariousness, when favored 131
defense of territory, when favored 125–7, 129, 131
definitions 125
optimal foraging strategy 126
resource competition, modes of 242–3
theoretical analysis, wild and provisioned groups 125–35, 147–50
Franje, female chimpanzee, Arnhem zoo 191, 198–9
Freud, S. 136
friendship
combined with hostility, provisioned groups 2, 144–6
Goodall and chimpanzee 176–7
human, same sex 202
wild chimpanzees 57
frustration
acceleration of state of 139
behavior without a goal, chimpanzees 140–1, 143–6
behavioral responses to 3, 67, 77, 89, 91, 100, 104, 136–7, 138, 139–42, 144, 182, 201
correlated with violence 69
defined 136
goal of 142
normally dormant state 142

frustration *cont.*
  prerequisite components of 31, 32–3, 136–7
  psychological view of 139
  responses of adolescent chimpanzees 119–20
  of restricted access to common goal 31
  see also aggressive behavior
frustration-aggression theory 136–42
  Barker, Dembo and Lewin experiment 101, 136, 137–9
  behavior without a goal 67, 69, 139–42, 182
functionalism 17

gatherer–hunters, *see* foragers
gene pool
  exclusion from 186
  widened through carnival gatherings 228, 229
genetic endowment of primates
  characteristic stimulus configurations 188
  equal potentials for aggressiveness or peacefulness 239, 260
  factor in charismatic–dependent relationship 252
  sets of behavior systems 149–50, 240
  variability of temperament 161–7
Gigi, female chimpanzee, Gombe 81, 82
goal(s)
  competitive 144
  cooperative 144
  defined 250
  of frustration behavior 139, 140
  individual, and group 250
  in mutual dependence system 250
  positive sanctions as 186
Goblin, male chimpanzee, Gombe 90–1, 93
Goliath, male chimpanzee, Gombe 27, 75, 76, 153–4, 183, 185, 246, 261, 265
Gombe National Park, Tanzania
    continuity of field studies in 196
    reduced size of 58, 255
Goodall, Jane xvii
Gray, female chimpanzee, Kibale forest 162, 200, 246
grooming
  achievement of adult rank through 186, 209–14
  an adult exchange 212
  charismatic chimpanzees 169, 208
  of despot apes 185, 208–9
  female–female 201
  functions of 209–10
  mother–child 202, 212
  non-reciprocated 210, 212–13, 214, 221–2, 263
  patterns, pre-1965 Gombe 210–12

Hadza, extant foraging society Africa 40, 41, 233
headmen, excluded from leadership theory 167
hierarchy, normal chimpanzee
  change without leader change 247–8
  changes form in crisis 247–8
  distorted form, Gombe and Malale groups 76, 248
  normally invisible 6, 245–6
  preference relations 245
  a social process 248
hierarchy
  in baboon groups 262
  in business firms 265
  dominance 34–5, 76, 167, 196, 244–5, 247
  form dependent on circumstances 6
  models of 6, 244, 247–8, 265
  theory 244–6
  see also dominance
home ranges, *see* territoriality
hunter–gatherers, *see* foragers

immediate-return system, *see* egalitarian societies
incest 264
independence, youthful
  advancing age of, Gombe chimpanzees 105–22
  distancing from mother 120, 206
  Mbuti forager children 259
  wild chimpanzee children 59, 107, 111, 113, 118, 206
infant and childhood experiences
  in foraging societies 43–9, 84, 87–8, 158
  of provisioned chimpanzees 28–30, 76, 77, 84, 88–95, 104, 106–14, 158–61
  in wild chimpanzee groups 84–8, 93, 94–5, 111, 112, 118, 131, 158, 159–61
infanticide 95
  ambivalent behavior of killers 97–8, 100–2
  explanations offered 101–2, 103, 130
  of infants of non-group females 100–1
  no attempt to stop killings 193
  redirected aggression 98
  within group 96, 99–100, 102, 130
  see also cannibalism
interference mutualism, *see* mutualism

Kahama, habituated chimpanzee group, Gombe 70

Kasakela, habituated chimpanzee group, Gombe 70
Kasonta, male chimpanzee, Mahale 169, 221–2, 224, 246, 247, 263
kinship
  categories considered kin 45
  cultural or social 45
  nature of, foraging society 45–6
  current views of kinship, chimpanzees 147–8
  social kin 45–6, 65, 229
  social fathers, chimpanzee 85–6, 90, 91–2, 156–61
  social parents, chimpanzee and foraging societies 86–8, 90, 91–2, 158–61
Kortlandt's hypothesis 51–2
!Kung San, extant foraging people, Africa 19, 40, 260

larger society gatherings
  basic unit of social organization 134, 260
  as foraging strategy 133–5
  a mammalian pattern 253
  network of local groups 43–4, 146, 260
  social kinship 45–6
  theory 231
  see also carnival reunions
leadership role
  of charismatic elders 234–37
  conferred by group 168, 171, 245
  defined 168
  fluidity of 46–7, 122, 168, 171–2, 234
  in foraging societies 46–8, 170–2
  theory of 167–8, 170
  among wild chimpanzees 154, 168–70
  see also charismatic leaders
Lorenz, Konrad xiv
Luit, male chimpanzee, Arnhem zoo 154, 155, 185, 199, 261

MacTavish, male chimpanzee, Guinea 235–6, 264
Male groups, *see* chimpanzees, and sedentary subgroups
Mama, female chimpanzee, Arnhem zoo 166, 198, 199, 200, 215, 236, 246, 261, 263
Man the Hunter symposium, 1965, 14–15, 19, 43
Mbuti pygmies, extant foraging people, Africa 19, 39, 64, 86, 161, 204, 235, 259
methods 1–35
Mike, male chimpanzee, Gombe 69, 75, 76, 77, 90–1, 164, 208, 246, 262
mobile and sedentary subgroups
  attachment of sedentary to home range 232–3
  a charismatic-dependent relationship 234
  chimpanzee 55–7, 63–4, 133, 206–8, 232
  demographic implications 232, 233–4
  of foragers 42–3, 57
  function of 133–4, 232–4
  Gombe patrols 146
  mutual dependence of 233–4
  structure of 51–7
Montagu A. 251
mother–infant relations
  baboons 259
  disturbed 89–90, 92–5, 101, 108–10, 111–12
  in foraging society 87, 110, 158
  grooming 212
  intimate bond 156–7
  maternal aggression 111–12
  weaning, normal 107, 108, 110, 111
  weaning, primates under stress 106, 107–12, 113, 259
mothering role
  aunting 215–16, 218
  generalized, foragers 87, 158, 175
  intimate, nurturing mother 85, 155–7, 175, 218
  learning 53, 215–16
  mutuality of parental roles 158
  precludes taking active leader role 53, 265
  strategy, protective retreat 157–8
  see also parental roles
mutual dependence relationship
  in aunting 217
  charismatic–dependent relationship 9, 155, 171, 174, 177, 178–9, 188, 250
  mobile and sedentary subgroups 233–4
  non-interference mutualism 178–9, 217
  parental roles 156–7, 158
mutual dependence system 3–4, 5–6, 8–11, 20, 151–93, 250
  adapted mode, chimpanzees 239
  mammalian 253
  model of 7–11, *10*
  principles of 9, 239, 241–2
  Rousseau's concept 249, 251
mutualism
  in foraging societies 47
  interference mutualism 178–9
  noninterference mutualism (NIM) 178–9, 217

natural selection 125, 126, 149–50, 240
New York Times xiv

'noble savage' tradition 20
norms
authority and 152
behavioral 179–80
of chimpanzees 161
conformity to 179–81, 184
core of culture 161
defined 180
deviance from 183
functions of 180
maintenance of 186
as means of social control 180–1
sanctions support 183

optimal foraging theory, *see* foraging theory
ostracism
among chimpanzees 181–6
definition, primate 183
inappropriate use of, Gombe group 182–3
in foraging societies 48
tacit use of, by human children 183
see also sanctions

Paliyam, extant foraging people, India 40
Pandarm, extant foraging people, India 40
Pandora, female chimpanzee, Guinea 175, 200, 235–6, 246, 262, 264
Pan troglodytes 4
paradigms
dominance 34, 35, 151–2
Kuhn on usage of 6, 34–5
resistance to change of 35, 256
tool of scientific method 15, 34–5
parental roles 155–61
in charismatic–dependent relationship 155, 174–8, 261
in foraging society 87–8, 158
generalized, between adults 155, 158–9, 175, 234–7
generalized, toward young 85, 86, 87, 90, 158–61, 175
mutual dependence of 158
stressed, Gombe group 90–3
see also fathering role *and* mothering role
Passion, female chimpanzee, Gombe 100, 262
patrols, Gombe chimpanzees 146–7
distortion of carnival 146–7, 230
mobile group 146, 147
negative excitement of 146, 147, 229–30
peace, state of
among animals 150
genetic potential for 149, 239
goal, maintenance of state of 250
need to study 153–4
not passive state 21, 48, 67, 152, 229
supreme value of foragers 18–20, 47–9, 131, 170
see also values and immediate-return societies
peace-maintaining role
of all adults 20, 154, 184, 199–200, 235–6, 263
of charismatic leaders 153–5, 235–7, 261
consensus of group 47, 199
normal role of dominant 76
socialization towards 47–8, 88
Pepe, male chimpanzee, Gombe 176, 208
positive behavior, *see* behavior, social
preferred partners
of chimpanzees 80, 120, 123, 162–5, 212, 213, 220, 221, 245, 263
preference relations 245
provisioning
catalyst for negative social change 3, 5, 241
effects on chimpanzees 5, 28–35, 59, 67–77, 79–84, 88–114, 116, 117–23, 127–34, 138–9, 141–2, 143–8
methods of Gombe 28–30
methods of Mahale 30–3
in overlap areas 127, 128
as schemata for creating frustration 29–30, 32–3, 69–70, 136, 137–9

rank order, *see* adult rank
reproductive success 95, 103, 155, 158, 254
role(s)
of average chimpanzee 167
authority inseparable from 152–3
charismatic leader 168, 170–1, 172, 234–7
chimpanzee awareness of 153, 154–5, 161
control, monkeys 154–5
defined 153
dependent follower 170–2, 190, 191–2
fluidity of 193, 234
theory, leadership 167–8, 169, 170
Rousseau, J.
enlightened self-direction 251
social contract 248–9

sanctions
in foraging societies 48–9, 184–6, 191, 261, 262
function of 48, 49

harmful deviance 190–1
negative 48, 181–2, 190–1
positive 48, 186–8
shunning, a passive mode of 182, 183–4
sociological understandings 48 9, 183
see also ostracism
scapegoating, *see* frustration
Schjelderup-Ebbe, T. 34
self-esteem
of charismatics and dependents 178, 250
core of healthy human personality 187–8
of foragers 48, 187–8, 250
positive social contacts build 187–8
sociopsychological goal of system 250
of wild chimpanzees 188, 250
self-interest
changed in form, provisioned chimpanzees 193
negative, selfish form of 250–1
positive enlightened form of 251
and self-esteem 251
sexual relations, chimpanzees
affected by social and psychobiological conditions 79, 82
in Arnhem zoo colony 219, 223–4
duration of copulation act 223, 257
evolved pattern of 80
in Gombe group 79–84, 220–1
in Mahale group 221–2
social form, carnival 79, 80, 222–3, 228
of wild chimpanzees 25, 77–9, 82, 83, 118, 186, 218–24
social bonds
affiliation theory 187, 231
charismatics and dependents 174–9
direct competition weakens 143, 144–5, 146–7
among foragers 45–6, 49, 131, 230–1
generalized male 202
to larger society 43, 44, 65, 70, 133–5, 146, 224–32
mobile and sedentary groups 232–4
not counter to adaptation 240
overemphasis on male 148, 202–3
weakened among provisioned chimpanzees 138
see also attachment to group
social climate
of foraging societies 18–19, 84
in provisioned chimpanzee groups 2, 67–70, 242
psychological use of term 57
in wild chimpanzee groups 2, 19, 57–60, 66–7, 84
social change
defined 242
in foraging society 14–15, 18–21
among free Kibale chimpanzees 257–9
in provisioned groups 3, 5, 8, 28–30, 33–5, 51, 59, 67, 68–77, 79–84, 89–114, 115–23, 138–9, 142, 143–9, 205, 241–2, 260, 264
substantive 242, 247
as survival mechanism 240
theory of 242–3, 247–8
social maturation
basis for acceptance to adult rank, chimpanzees 206–9
basis for acceptance to adult rank, foraging society 203–5
delayed, of Gombe young 105–14, 117–22
socialization theory 84, 88
status 152–3, 173
structure of attention, *see* attention structure
Strum, S. 259

temperament
affects role taken 190, 191–3
difference of, charismatic-dependent partners, chimpanzees 123, 162–5
variability of, primates 155, 161–7, 190
territoriality
definition of, anthropological 60
of foragers 42–5, 60–1, 135
foraging theory 125–7, 129, 131, 134, 242–3
optimal foraging theory 126
of provisioned groups 60–1, 70–4, 105, 117, 122–3, 127–31, 134–5, 147–8
of wild chimpanzees 44–5, 60–7, 70, 114–15, 127–8, 133–5, 260

Vayda, A. P. 72
values, of immediate-return societies
avoidance of conflict 131
essential structural and functional properties 20, 41, 47–9
of foragers 18, 20, 131, 170
positive social relations 20, 41, 45–6, 131
practical, of chimpanzees 131, 249
Western society 172, 250, 251
see also peace, state of
Virchow, R. xiii
visiting
by foragers 43, 44, 46, 132, 227
of wild chimpanzees 106, 114–17, 211, 228, 261
see also carnival reunions

Western society
  adolescence in 203
  negative view of dependence 172, 256
  perception of self-interest 250–1
  values 172, 250, 251
William, male chimpanzee, Gombe 27, 163, 164, 205

Yeroen, male chimpanzee, Arnhem zoo 154, 185, 199, 200, 224, 246

youth, period of
  in foraging society 117, 118, 203–5, 224, 259
  Mahale males accept youthful 208
  wild chimpanzees, 117–22 *passim* 205–24
  see also adolescence, period of

Zira, female chimpanzee, Kibale forest 162, 191, 192
Zuckerman, S. xiv